The Basic Soldering Guide Handbook

Alan Winstanley

In association with

Written, designed and published by A R Winstanley

All text, diagrams and photographs are fully protected by copyright and may not be reproduced in any commercial or non-commercial publication or medium without the prior permission of the writer.

Every care has been taken to ensure that the information and guidance given is accurate and reliable, but since conditions of use are beyond the writer's control no legal liability or consequential claims will be accepted for any errors contained herein.

Where mentioned, the U.K. mains voltage supply is 230V a.c. and you should amend ratings for your local conditions.

Copyright © 1996- 2014 A.R. Winstanley

All Rights Reserved.
First published in Amazon Kindle® format July 2013
This paperback edition published August 2014

Photography Sony α DSLR 50mm macro
Printed by Createspace, an Amazon.com company

www.alanwinstanley.com

ISBN-10: 1500531146
ISBN-13: 978-1500531140

The Basic Soldering Guide Handbook

CONTENTS

1	Introduction	6
2	First Steps	8
3	Preparation	17
4	Solder & Fluxes	28
5	Soldering Step by Step	40
6	Reflow Soldering & Desoldering Techniques	63
7	Useful Checklists	77
	Troubleshooting Guide	79
	Potential Hazards	80
	Basic First Aid	81
	Useful Resources & Web Links	82
	Conclusion	83

Dedicated to Mum, Dad and Robert for encouraging my schoolboy electronics tinkering in the pioneering days of hobby electronics, and to everyone at EPE Magazine past, present or absent for their support over many years.

ACKNOWLEDGMENTS

Antex (Electronics) UK generously provided samples of their ever popular British-made soldering equipment, materials and accessories for use in the Basic Soldering Guide Handbook. When starting out as a 1970's teenage schoolboy electronics hobbyist, I and my trusty Antex soldering iron shared many adventures in electronics together, and it has been a pleasure to use the latest Antex soldering irons throughout this guide.

Also I wish to thank Brian Brooks of Magenta Electronics Ltd. (www.magenta2000.co.uk) for supplying a generous variety of their professional printed circuit boards and a range of components which were the subjects used in the photographs.

PREFACE

In 1996 when the world wide web was very young, I launched the first and most detailed website ever describing the practical skills of electronic soldering, and my *Basic Soldering Guide* quickly became the No. 1 web site of its kind in Google. Thanks to its in-depth reference text and the unequalled high-quality close-up photographs showing soldering step by step, many quickly learned the essential stages needed to make a solder joint successfully. Even novices who had never tried soldering before, gained the skills and confidence needed to acquire this skill.

My *Basic Soldering Guide* became a key go-to online guide for soldering, and I've enjoyed receiving encouraging feedback ever since from the likes of the US Air Force, US Marines, US Coastguard, Honeywell trainers, Atomic Energy authorities, Australian aeronautical suppliers, UK colleges and universities, trainees and many more around the world.

In association with Antex (Electronics) Ltd., the leading UK manufacturer of electronic soldering equipment, I'm delighted to bring you this Handbook edition of my *Basic Soldering Guide* containing over 80 all-new colour photographs, more background, more detailed information and lots more practical hints and tips.

I've taken onboard readers' queries and some 17 years of online feedback, experiences and answering reader's questions. With the help of all-new photography, I'm sure you'll master the skills needed to solder electronics successfully using this updated guide.

I welcome feedback and comments, and readers can reach me via my website at www.alanwinstanley.com or Email me at alan@epemag.demon.co.uk.

Alan Winstanley   United Kingdom, August 2014

1 INTRODUCTION

The first and most important aspect of assembling any electronic project is that of **soldering**, which is a delicate and precise skill that can be mastered with experience. Sometimes called "soft soldering", there's no shortcut to acquiring the necessary expertise, and producing a consistently satisfactory solder joint takes a little practice. However, like riding a bicycle, soldering is an art which once learnt is never forgotten, and the purpose of this new and updated guide is to explain the techniques of soldering and desoldering for beginners, which I hope will set the hobbyist or trainee technician firmly on the road to successful electronic assembly or repairs in the future.

• **Soldering** is the least "aggressive" way of joining non-ferrous metals together, and is used universally in electronics, air conditioning and refrigeration circuits, household plumbing and more besides – applications where the precise joining together of components at fairly moderate temperatures is needed.

• Further up the scale, **brazing** involves using higher temperatures to melt brazing rods onto larger metal parts, perhaps to repair a metal chair, lawnmower or to fabricate metal components or jewellery into intricate shapes.

• Lastly, **welding** is a very aggressive way of fabrication using welding rods or wire; steel girders, oil rigs and ships are all welded together, or robotic spot-welding is used for the mass production of, say, washing machines or car bodyshells using sheet steel to make strong rigid assemblies.

Due to the lower temperatures used and the need to make consistently good electrically conductive and mechanically sound joints with precision, soldering is used to connect components in manufacturing electronic circuits. Small components would quickly be destroyed by brazing or welding, although tiny spot-welding joints do appear in electronics, perhaps to weld a metal tag onto a button battery.

The Basic Soldering Guide Handbook

This guide therefore deals with the soldering techniques used in electronics at hobbyist or trainee educational level. It explains what to look for before buying a soldering iron, describes ways of making various solder joints on circuit boards and other electronic components, and also how to desolder – removing solder in order to repair a circuit board or replace an electronic component.

You'll also find more details of other aspects of soldering, including an outline of typical solder types and fluxes. In short, everything you need to get started in electronics soldering is here, so let's get started!

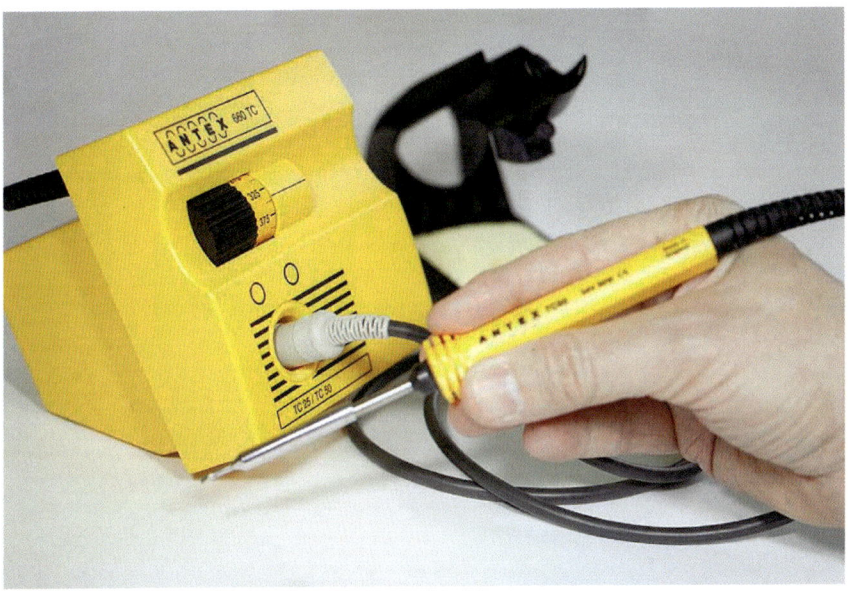

For enthusiasts or industry, this Antex 660TC soldering station has a separate mains-powered control unit and a matching low-voltage soldering iron rated at 24 Volts, 50 Watts so it's more suited to a wider range of tasks than a lower powered one.

2 FIRST STEPS

The principle behind soldering sounds quite simple: the idea is to join components together to form an electrical connection, by using a mixture of lead and tin solder or alternatively "lead-free" solder (an alloy of tin and copper), which is melted onto the joint using a soldering iron. If you have never picked up a soldering iron before, then this guide will show you everything to help you start soldering with confidence. I also hope that the guidance will help those working in other areas — computer technicians or audio enthusiasts, for example, who may be faced with electronic repairs or modifications using a soldering iron for the first time.

If you're an electronics hobbyist or trainee, before embarking on any form of ambitious electronic project, it is recommended that you practice your soldering technique on some brand new components using clean strip board (or protoboard) or a printed circuit board, and select a simple and straightforward constructional design as a starting point. Become acquainted and comfortable with your chosen soldering iron, which likely to become as familiar to you as a favourite pen. Learn how to balance it and handle it with precision. Try soldering an assortment of resistors, capacitors, diodes, transistors and integrated circuits with it, and then try your hand at desoldering – removing the solder again to make a repair or modification.

A really good place to start learning is by building a simple electronics kit, such as a Velleman kit that contains a good quality printed circuit board. You'll learn some of the basic skills of successful soldering this way, and it'll be a great confidence booster too.

Did you know? In the USA and elsewhere, the letter L in "solder" is silent and they say "soda" or "sodder" – but here in Britain we do pronounce the L and we say "sole-der"!

Search any electronics catalogue or website and you'll see a bewildering array of soldering equipment on sale, including irons, controllers, work stations and desoldering equipment too. A large range of soldering irons is readily obtainable - which one is suitable for you depends on your budget and how serious your interest in electronics is, but there's something for every pocket distributed by a variety of retail, industrial and mail-order outlets.

The Antex range of soldering equipment has been very popular with industry, education and the electronics hobbyist for 60 years. An industrial user or a more dedicated hobbyist – with a bigger budget! – will be interested in a soldering station instead and again Antex offers a range of British-made products for industry or home use.

A very basic mains electric soldering iron can cost from under £5 (US$ 8), but I find that these very cheap irons, as sold on auction websites, are pretty crude and imprecise. They are best suited for simple electrical repairs and DIY rather than precision electronics or printed circuit boards discussed here. They tend to be bulky and uncomfortable for extended use, and they may not have suitable "bits" or tips of various sizes to suit different tasks.

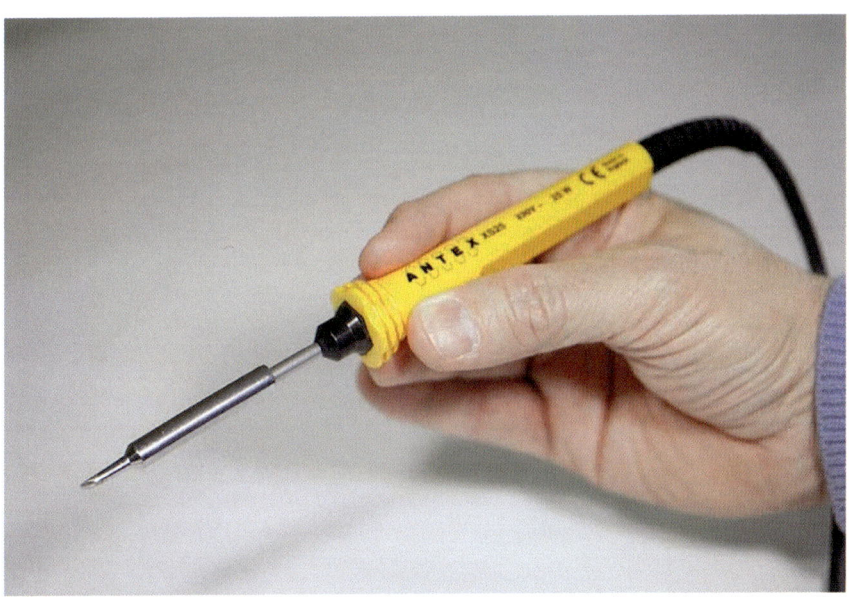

This classic Antex XS25 "pencil style" 25W mains-powered soldering iron has exchangeable tips or "bits" and is ideal for hobbyists and students.

A quality pencil-style electric soldering iron such as the Antex XS25 (previous photo) will be approximately £20 (US$18 tax free) - though it's possible to spend into three figures on a soldering iron "station" if you're really serious about the subject.

Don't be tempted to over-spend on an elaborate workstation though, unless you are really very serious about becoming involved in electronics. You will usually obtain perfectly satisfactory results using a fairly modest "pencil" iron model, and you can upgrade to something more sophisticated should your needs change in the future.

When choosing your soldering iron, certain factors to bear in mind include:

Voltage: for the British market, "mains" electric irons run directly from the mains at 230V a.c. or will be set for other voltages (110V a.c.) depending on the country. However, low voltage types (e.g. 12V or 24V) usually form part of a "soldering station" for use with a matching controller made by the same manufacturer. Some low-voltage irons run off batteries (e.g. a car battery or Ni-cads) but these are uncommon.

Wattage: this is an extremely important factor to think about when choosing your iron. Typically, irons for general electronics work may have a power rating of between 15-25 watts or so, which is fine for most electronic assembly tasks, printed circuit boards and inter-wiring. It's important to note that a higher wattage does not mean that the iron runs hotter - it simply means that there is more power "in reserve" for coping with larger joints. The maximum electric iron wattage generally available is about 100W, which is OK for DIY electrical repairs but is far too high for general electronics or circuit board use.

A higher wattage iron offers you more flexibility for tackling a wider range of tasks. It has a better "recovery rate" which makes it more "unstoppable" when it comes to heavier-duty work, because it won't be drained of its heat so quickly. So check the power ratings carefully, and anything between 15–40W is fine for general electronics soldering.

Temperature Control: the simplest and cheapest types don't have any form of temperature "regulation". Simply plug them into the mains and switch them on! Thermal regulation is "designed in" (by physics, not electronics!). Sometimes they are described as "thermally balanced" as they have some degree of temperature "matching" – in other words, they warm up as quickly as they lose heat during use, so in a primitive way they maintain roughly a constant temperature. This type of iron is perfectly acceptable for hobby or less demanding professional use. It's also essential to use the manufacturer's specified tips (see later) to maintain proper temperature matching, otherwise the iron may not heat up enough – or it may overshoot in temperature, due to incorrect thermal matching of major components.

These unregulated irons form an ideal general-purpose iron for most users, and they cope reasonably well with printed circuit board soldering and general interwiring. However, most of these "miniature" types of iron will be of little use when attempting to solder large joints (e.g. very large terminals or very thick copper wires) because the components being soldered will draw or "sink" heat away from the tip of the iron, cooling it down too much and preventing solder from flowing properly. That's when a higher wattage iron is needed.

A proper temperature-controlled iron will be quite a lot more expensive - retailing at say £40 (US$ 60) or more - and will have some form of built-in thermostatic control, to ensure that the temperature of the "bit" (the tip of the iron) is maintained at a fixed level within reasonable limits. This is desirable during frequent use, since it helps to ensure that the temperature will be relatively stable regardless of the workload. Some irons have a bimetallic strip thermostat built into the handle which gives an audible "click" in use, and some may include an adjustable screwdriver control within the handle as well. Others may have electronic controls built in.

More expensive still, **soldering stations** cost from £70 (US$ 115) upwards (the iron may be sold separately, so you can pick the type you prefer), and consist of a complete bench-top control unit into which a special low-voltage soldering iron is plugged. Some versions might have a built-in digital temperature readout, and a control knob to vary the setting. The temperature could be boosted for soldering larger joints, for example, or for using higher melting-point solders (e.g. lead-free or silver solder). These are designed for the discerning user, or for continuous production line or professional use. A thermocouple will be built into the tip or shaft, which monitors temperature.

The best soldering stations have irons which are well balanced, with comfort-grip handles which remain cool all day, and silicone-based cables which are burn proof. Antex produces a range of irons with silicone cables specially for education use, to help avoid accidents caused by careless use by students.

Anti-static protection: if you need to solder a lot of static-sensitive parts (e.g. CMOS chips or MOSFET transistors), more advanced and expensive soldering iron stations use static-dissipative materials in their construction to avoid static charges accumulating on the iron itself, which could otherwise damage or "zap" some semiconductors. Such irons are listed as "ESD safe" (electro-static discharge proof).

The cheapest irons are not classed as ESD-safe but they still perform well enough in most hobby or educational applications provided you take the usual anti-static precautions when handling these components. The iron would need to be well earthed (grounded) in these circumstances, to carry away any static.

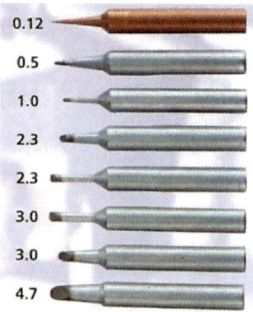

A range of spare tips or "bits" produced by Antex for their soldering irons. A 3mm type is fine for general electronics work.

Tips or Bits: it's often useful to have a small selection of manufacturer's bits (soldering iron tips) available with different diameters or shapes, which can be changed depending on the type of work in hand. You will probably find that you become accustomed to, and work best with, one particular shape of tip for the majority of your work. I mainly use a 3mm size for general soldering. Usually, tips are iron-coated or plated to preserve their life and to maintain good tip "hygiene". Only use tips designed for your specific model of soldering iron, or thermal problems may arise.

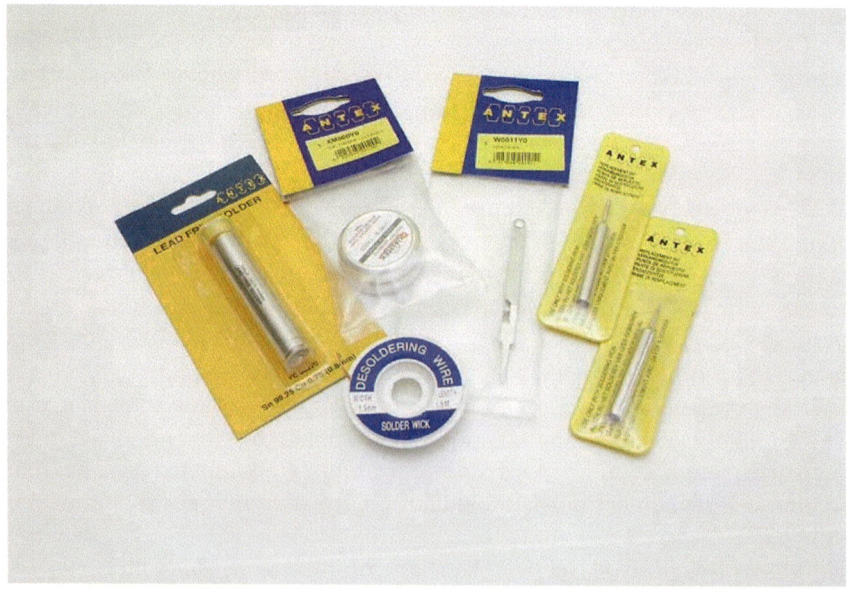

A range of soldering accessories produced by Antex.

Spare parts: it's always reassuring to know that spares are available in the future if required, so if the element blows, you don't need to replace the entire iron. This is especially the case with expensive irons. Check some websites or mail-order catalogues to see whether spare parts are listed. One drawback is that you may need to use another soldering iron when exchanging a broken heating element!

Gascat 40 butane soldering iron kit by Antex.

Gas or electric? So far I've discussed electric soldering irons, but gas-powered soldering irons are sold which use butane propellant rather than mains electricity to operate. They have a flint lighter or (better still) a built-in piezo for ignition, and have a catalytic element which, once warmed up, continues to glow hot when gas passes over them. They tend to be big and bulky compared to electric pencil irons.

Field service engineers use gas-powered irons for working on repairs where there may be no power available, or where a joint is tricky to reach with a normal electric iron, so they are really for occasional "on the spot" use, rather than for mainstream construction or assembly work. I use one for quick repairs when I can't be bothered getting the electric soldering iron going!

Gas irons can have higher power equivalents than electric ones (eg 125 watts or more) but some gas-powered irons are nothing more than miniaturised blowtorches, which may or may not be useful for occasional heavier duty soldering. Almost every electronics constructor uses an electric-powered iron for general use. If you are considering buying a gas soldering iron then the author's book "*An Introduction to Gas Soldering Irons*" (Kindle) explains their operation.

A **solder gun** is a pistol-shaped iron, typically running at 100W or more, and is completely unsuitable for soldering modern electronic components: they're too hot, heavy and unwieldy for micro-electronics use, nor are they designed for that. Plumbing or DIY, maybe..!

Above, a basic heat-resistant soldering iron stand with cellulose cleaning sponge. (Courtesy Antex (Electronics) Ltd.)

A benchtop fume extractor fan for hobbyist use. A replaceable carbon filter helps remove particles and air is vented out the back.

The Basic Soldering Guide Handbook

Soldering irons are best used along with a heat-resistant bench stand, where the hot iron can be safely stored in between use (photo opposite). It is extremely important that a hot soldering iron is safely "parked" ready for action, and a bench stand is really a necessity. Soldering stations usually have such a feature, otherwise a separate soldering iron stand is essential, ideally one that's supplied with a tip-cleaning sponge. You can make your own cleaning sponges using cellulose sponge only.

Other equipment worth considering includes the use of fume extractors, which are compulsory in the industrial workplace. Solder fumes and flux smoke are not known to be toxic, but they can cause irritation.

A basic fume extractor (photo opposite) consists of a small bench-top fan which draw fumes and irritating smoke away from the operator's face and filters out some of the smoke particles, before exhausting the air back out through the fan. The carbon-impregnated foam filters are replaceable. Such devices are quite effective and users soon find them indispensable, even though they can be a bit noisy at close range.

Professional fume extraction systems draw the smoke and fumes directly from the work area via a clip-on tube fitted to the soldering iron, then vent the fumes away through a large filter pump. It is definitely worth considering a small bench top unit for regular hobby or occasional professional use, as having decent ventilation can only be a good thing.

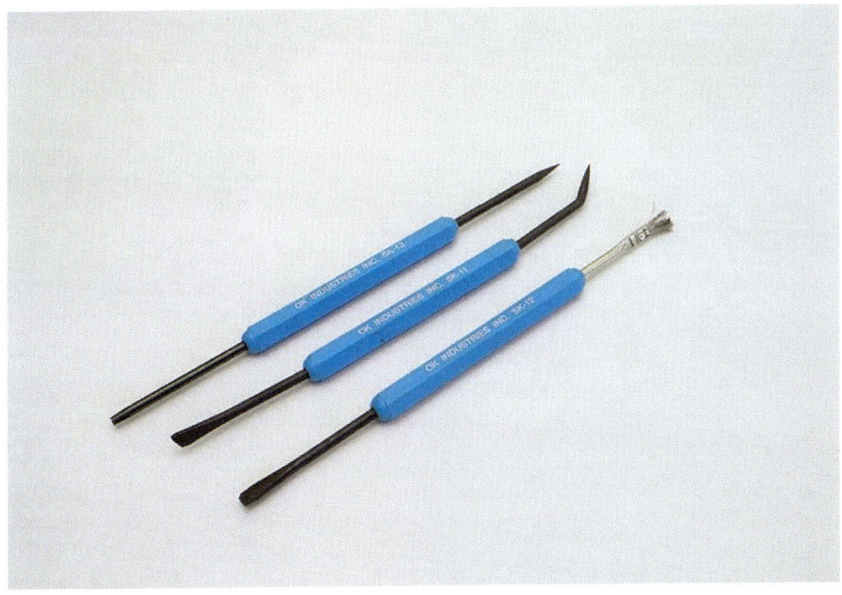

A soldering accessory toolkit by OK Industries, including probes, scrapers and wire brush.

A variety of specialist hand tools are available that assist with soldering, and a good supplier's catalogue will offer a range of small brushes, scrapers and cleaning tools (shown in the preceding photo) in a handy kit, together with the usual types of wire cutters, pliers and so forth, which are necessary for handling components and tidying up as required. Some specialist service aids (aerosols etc.) are also described later.

At the end of the *Basic Soldering Guide Handbook* I have included a list of soldering items and tools that I classed as Essential, or Good to Have, or "Luxury" items that advanced dedicated users might appreciate..

Now let's look at how to use soldering irons properly, and later on we will describe the techniques for putting things right when a joint somehow goes wrong — and don't worry, even the experts get it wrong sometimes!

3 PREPARATION

This guide will show you step by step how to solder successfully and plenty of photographs are provided to help explain the techniques needed. As you read through the guide I'll explain all the stages in more detail, but let's look at the basics first.

First of all, successful soldering requires that the items being soldered together are held with as little movement as possible. So it's best to secure the work as needed, so that your accuracy isn't affected should the workpiece move accidentally.

In the case of a printed circuit board, various holding frames are useful especially when densely populated boards are being soldered: the idea is to insert all the parts on one side (a process called "stuffing the board"), hold them in place with a suitable pad to prevent them falling out again, turn the board over and then snip off the wires with cutters before soldering all the solder joints in one go. The frame saves an awful lot of turning the board over and back again, especially with large boards, as all the soldering can be performed in one pass.

Only the more serious constructor will purchase a holding frame, and hobbyists can retain parts in place in other ways – the popular "Helping Hands" gadgets cost a few pounds and have crocodile clips to grip parts, and maybe a magnifying glass or soldering iron stand too. The cast iron base provides stability. Other parts could be held firm in a modeller's small vice.

When soldering parts onto an ordinary circuit board – known as through-hole soldering – components' wires can simply be bent to the correct pitch (distance apart) to fit through the board, insert the part flush down against the board's surface, splay the wires a little so that the component grips the board under spring tension, and then solder it. This technique is universally used in manual through-hole soldering, which is explained in full later.

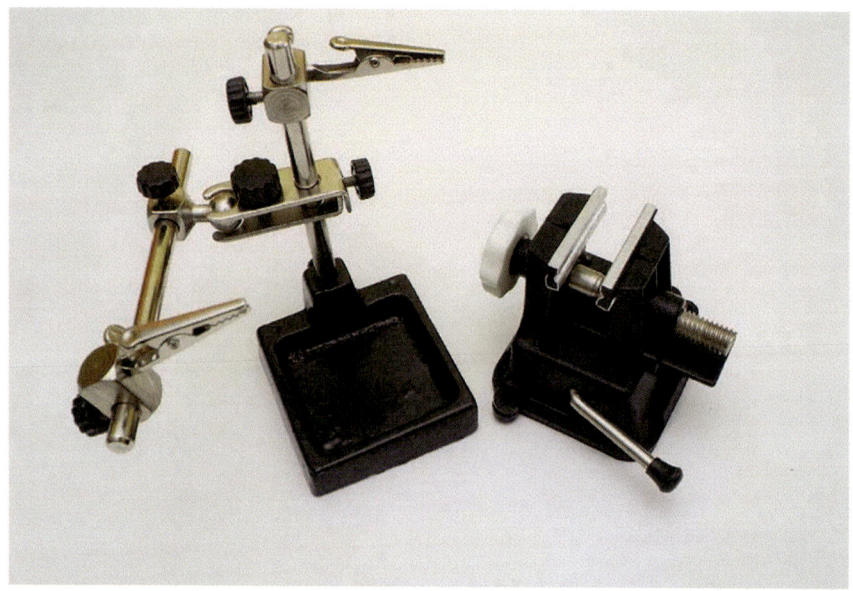

Above, the ever-popular "Helping Hands" (left) helps support sundry parts, wires etc. during soldering. A modeller's vice (right) holds parts firmly. A vacuum base fixes it onto smooth surfaces.

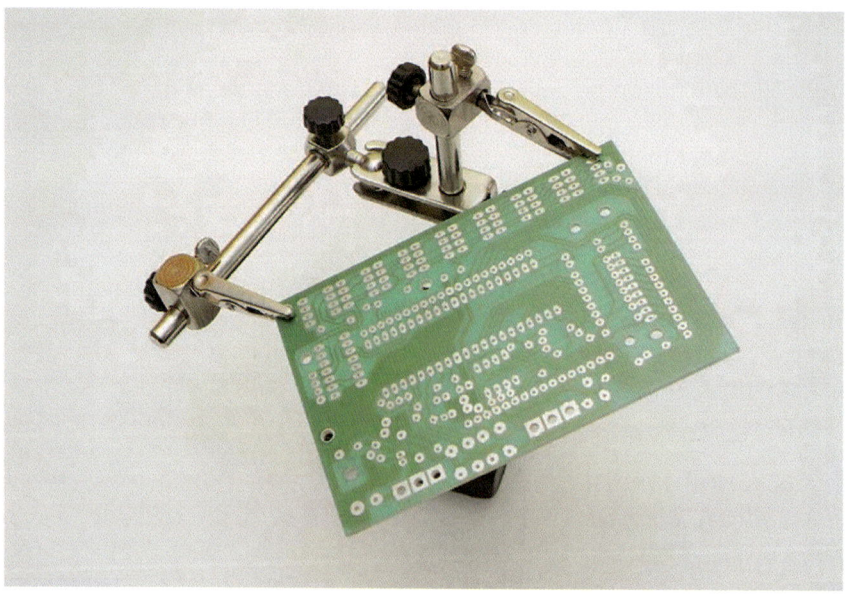

"Helping Hands" uses crocodile clips to grip parts during soldering. Or just place boards flat on the workbench.

In the author's view - opinions vary – it's generally better to snip off the surplus wires leads first to make the joint and neighbouring joints more accessible and also to reduce the mechanical shock transmitted to the p.c.b. copper foil. However, in the case of diodes and transistors the author tends to leave the snipping until after the joint has been made, since the excess wire will help to "sink" heat away from the heat-sensitive semiconductor junction.

A special clip-on heatsink is available which also helps stop excess heat from reaching temperature-sensitive semiconductors like these. I've always managed without one but beginners might find them reassuring. Integrated circuits can either be soldered directly into place if you are confident enough, or better, use a dual-in-line socket to prevent heat damage. The chip can then be swapped out at a later date if needed.

Parts which become hot in operation (e.g. some power resistors), should be raised above the board slightly to allow air to circulate. Some components, especially large electrolytic capacitors, may require a mounting clip to be screwed down to the board first, otherwise the part may eventually break off due to vibration. It's a good idea to bolt such components firmly into place before soldering their terminals, in order not to strain the soldered joints or the components when fasteners are tightened. By securing the workpiece as much as possible to prevent movement, you have a much higher chance of producing good-quality reliable solder joints.

Let's get to grips with actual soldering techniques in more detail. The soldering of electronics components utilises lead/tin or lead-free solder and the process is compatible with many non-ferrous metals. You can solder copper, lead, brass, gold plate, silver, nickel, tin and tin plate, zinc and more besides but some metals such as nichrome, galvanised or stainless steel require a highly specialist "flux" (see later) to solder them and aren't discussed here. Some materials such as beryllium, chromium, magnesium and titanium are non-solderable in any case, according to solder manufacturers Multicore.

In electronics we're mainly concerned with soldering parts or wires onto printed circuit boards or terminals that are usually already coated or "tinned" with solder or plated, ready for soldering with flux-cored solder.

The key factors affecting the quality of a solder joint are:

• **Cleanliness** – dirt or impurities drastically hinder good solder coverage.
• **Temperature** – the right level to enable the solder to flow freely!
• **Time** – apply heat for just the right amount of time!
• **Adequate solder coverage** – enough to form a good joint without touching neighbouring areas.

A little effort spent now in soldering the perfect joint may save you — or somebody else —a considerable amount of time in troubleshooting a defective joint in the future.

Let's discuss the basic principles outlined above in more depth. Firstly, and without exception, all parts - including the soldering iron's tip itself - must be clean and free from contamination. Fact is, solder just will not "take" to dirty parts! Old components or copper board can be impossible to solder because of oxidation that builds up on the surface of the leads.

Impurities repel the molten solder and this will soon become evident because the solder will "bead" into shiny globules looking like mercury, going everywhere except where you need it. Dirt and contamination are the enemies of a good quality soldered joint!

A clean and shiny soldering tip or "bit" is essential for each and every solder joint you make.

When all the conditions are right for soldering, materials are said to be "wettable" and will accept molten solder, which should flow readily over their surfaces. Hence, it's really necessary to ensure that parts are free from grease, oxidation and other contaminants.

Note that in any case, some incompatible materials or surface finishes just cannot be soldered using ordinary lead-free solder, no matter how hard you try – e.g. aluminium parts would require special aluminium solder and fluxes (see later) to be used.

The Basic Soldering Guide Handbook

The two tags on this switched potentiometer have blackened with age and oxidation. They must be cleaned before they can be soldered.

In the case of old components that have been stored a long time, where the leads have blackened and oxidised, use a small hand-held file or perhaps fine emery cloth to reveal fresh metal underneath. Stripboard and copper printed circuit board will generally oxidise after a few months, especially where it has tarnished due to fingerprints, so the copper strips could be cleaned using e.g. an abrasive rubber block made for the purpose.

Also available is a fibre-glass filament brush (photo, next page), which is used propelling-pencil-like to remove contamination. They're OK for general contact cleaning but are best avoided on fine surfaces (e.g. gold plated switch contacts). They also produce tiny particles which can irritate the skin, so avoid accidental contact with any debris.

Above, a glass-fibre filament brush like this is useful for cleaning oxidised parts. Refills are often available.

The glass-fibre brush works like a propelling pencil and produces irritating dust. Avoid skin contact with debris as much as possible.

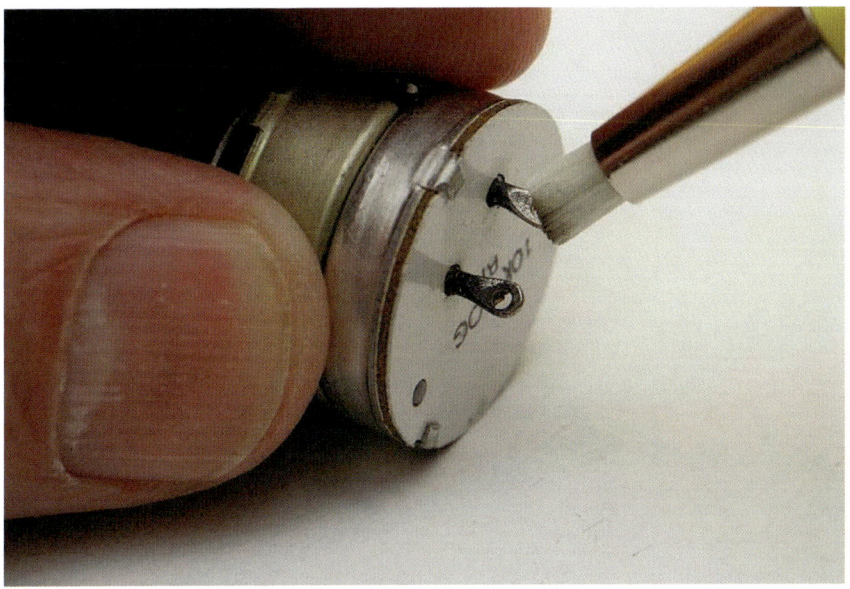

Cleaning oxidised tags with a glass fibre brush to make them nice and shiny, ready for soldering.

After cleaning, a wipe with a cloth soaked in suitable solvent such as Isopropanol will remove most grease marks and fingerprints. After preparing the surfaces, avoid touching any parts as far as possible.

Another side effect of trying to solder contaminated or incompatible materials is the tendency to apply more heat and "force the solder to take". As the materials aren't wettable the molten solder won't flow where you want it to flow. This can do much more harm than good, because it may be impossible to burn off any contaminants anyway, and the component or the printed circuit board may be overheated and damaged in the process.

Semiconductors may be harmed by applying excessive heat for more than a few seconds, and extreme heat applied to printed circuit board tracks can cause irreparable damage, because the tracks will be lifted away from the substrate especially on a delicate or badly designed board. (This problem is highlighted at the end of the book.) You can avoid trouble by ensuring the surfaces to be soldered are clean and wettable to begin with.

As already explained, cleanliness in soldering is a major factor in obtaining successful results. A soldering iron stand usually has a sponge that is dampened – it wants to be quite damp but not running wet, so squeeze out any excess.

Above, dampen the sponge but don't soak it unduly. The hole in the sponge gives operators an edge to wipe the soldering iron tip.

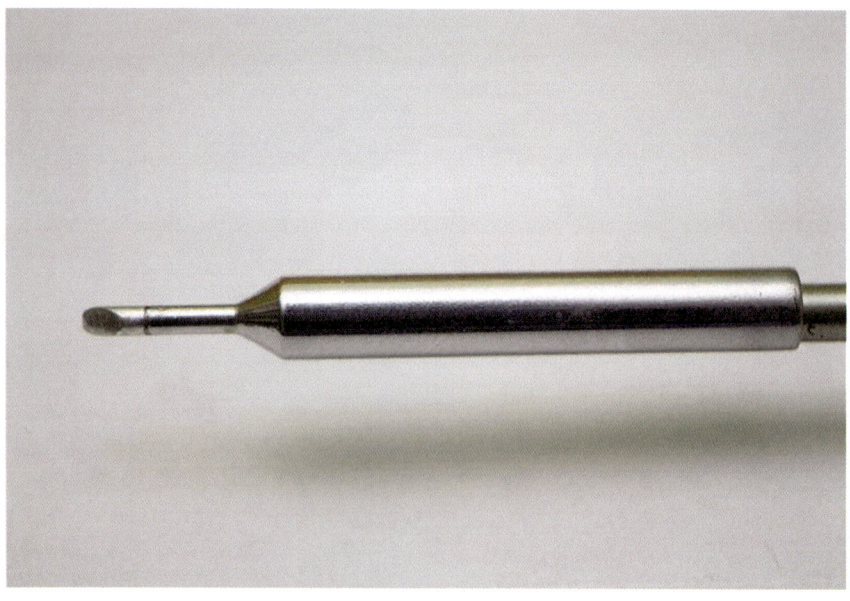

An Antex soldering iron tip. This model simply slides onto the iron's heating element. Only the extreme end is "wettable" and will accept solder: the tip should be kept nice and shiny.

Soldering iron bits are typically iron-plated to resist wear, then chrome-plated to prevent molten solder being deposited on them, with only the extreme tip being "wettable" to work with molten solder. Before using the iron to make a joint, the hot tip must be "tinned" with a few millimetres of solder: you should always flood a brand new tip with plenty of solder to tin it immediately, when using it for the first time.

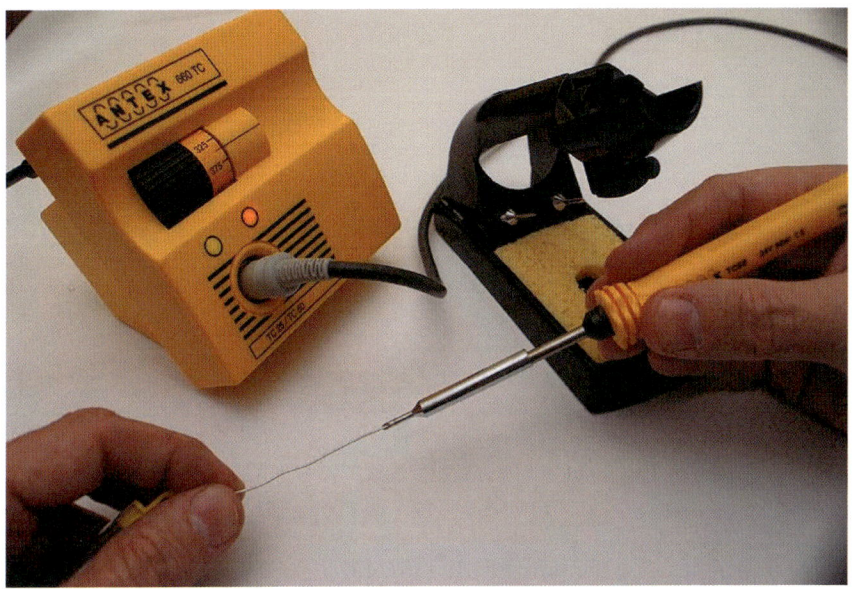

Tinning the tip ready for use: apply a few millimetres of solder to the hot tip then wipe off any excess on the damp sponge. New soldering tips must be flooded with solder straight away before use.

Wipe off excess solder using the damp sponge and it's then ready to use. That's why sponges have a hole or well in them – the edge acts as a wiper and the hole catches any excess deposits of solder.

Just before using the iron it helps to re-apply a small amount of solder to a clean tip, to improve the thermal contact between the tip and the joint. The molten solder fills the void between the materials being soldered and the iron tip, to help transfer heat better so that the solder flows more quickly and easily.

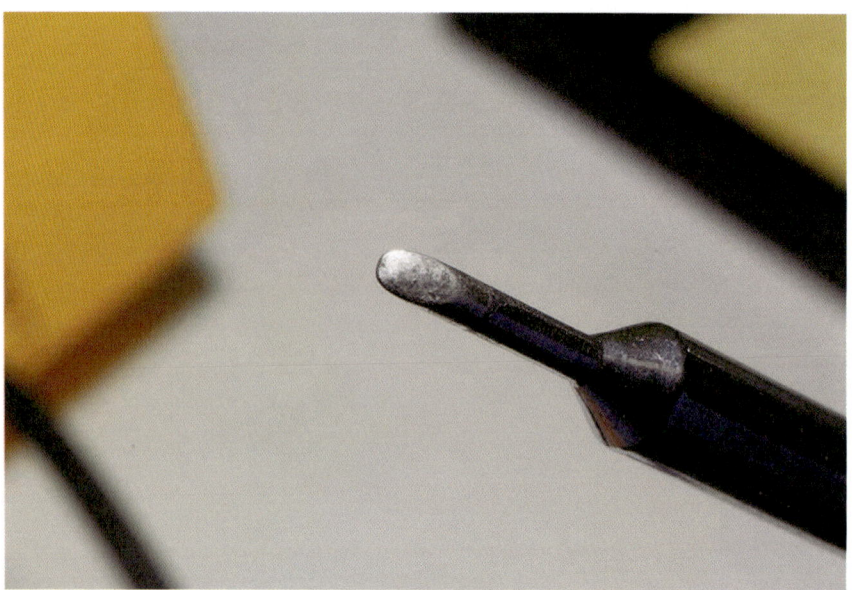

Above, a tinned tip should look like this – clean and shiny with no contaminants evident, ready to make the next joint.

Multicore TTC1 Tip Tinner and Cleaner is a useful aid to maintaining a soldering iron bit in good clean condition.

It's sometimes better to coat or "tin" larger parts with molten solder as well before making the joint itself, but this is not generally necessary with p.c.b. work. Multicore Tip Tinner & Cleaner is very useful: gently abrasive in action, it helps to clean dirty bits and keep them in good condition. Use them for removing stubborn contaminants, but don't overdo it.

Just press the hot iron onto the solid paste and scrub it around a little. The tip will be cleaned, tinned, and made ready for use.

The move to lead-free solders (see next section) has had some effect on the life of soldering iron bits, with increased wear and corrosion noted due to the higher temperatures and the fluxes found in tin-based solders. You can therefore expect bits to wear out over time.

Once the iron-plating is damaged due to oxidation or erosion, the bit is fast approaching its end of life. Never use an abrasive or file to sand down a tip: the iron-coating will be damaged and the iron's core exposed, so the tip will soon be made useless due to erosion.

In the next chapter, we look at solder and fluxes – essential ingredients of a successful solder joint.

4 SOLDER AND FLUXES

Having prepared the soldering iron tip ready for use, let's turn to the key subject of solder and fluxes. In recent years there's been a move towards using more environmentally-friendly materials in electronic products. EU legislation such as the *Restriction of the Use of Certain Hazardous Substances in Electrical and Electronic Equipment* (RoHS) aims to reduce toxic heavy metals being sent to landfill. (Look for the RoHS symbol on equipment to indicate compliance.) Due to RoHS compliance, the electronics industry had to change the type of solder it uses in electronic production.

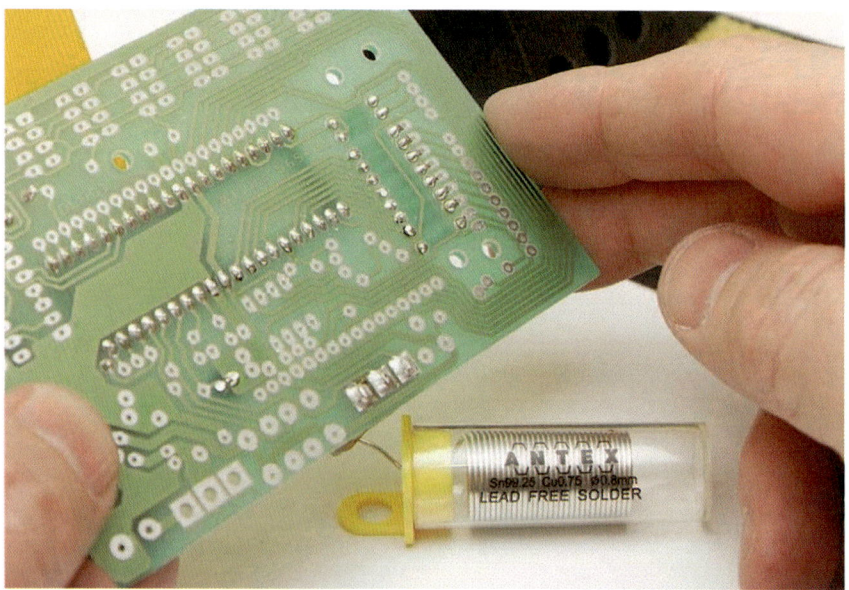

Electronics solder is supplied in reels or dispensers like this.

Solder comes in various forms including solid bars or pellets for melting in small electric 'solder pots' used for treating the ends of wires with solder. Traditional general-purpose electronics-grade solder is in wire form – starting with so-called "60/40" which contains 60% tin (symbol Sn) and 40% lead (symbol Pb) and is sold in handy dispensers or reels. Although tin-lead solder is now banned in industry, there's nothing to stop the hobbyist from using it but best practice is to use lead-free solder in our work: my advice is to try both, and see which you prefer to work with. "40/60" tin-lead produces lower quality results but is slightly cheaper and perfectly acceptable in hobby circles.

Various diameters of solder wire are marketed. In the UK they're sold in Standard Width Gauge (SWG) sizes, typically as 18SWG (1.2mm) or 22SWG (0.7mm)*. The latter is fine for almost all hand-soldering of printed circuit boards or general electronics. For larger solder joints (e.g. larger switch or motor terminals), 18SWG solder would be better as more solder can be dispensed more quickly.

* In the USA, American Wire Gauge (AWG) is specified, and a typical general purpose solder is 21AWG /0.032".

Lead-free solder is universally available and contains typically 99.7% pure tin and 0.3% copper (symbol Cu). It needs a higher melting point which makes it slightly more difficult to work with, but standard soldering irons will cope with it well. Antex lead-free solder (Sn 99.25 / Cu 0.75) is a good compromise at 0.8mm diameter (equivalent to 21AWG) and is sold in small dispensers.

Other solders are produced for specialist work, including aluminium solder (Alu-Sol®) and another solder variant used by professionals is Multicore "Smart" wire which contains a small amount of pure silver (symbol Ag). It produces very clean results and is often associated with SMD (surface mount devices), though some engineers also use it for routine p.c.b. work for producing the best possible finish by hand. As "Smart" wire contains lead it is not RoHS compliant.

An interesting variant is Eutectic solder, which is 63/37 Tin/Lead. It goes instantly from solid to liquid when melted and is particularly good for hand-soldering. An almost-equivalent lead-free product would be Stannol Flowtin TC or TSC solder.

When melting solder with a soldering iron, oxides of metal are produced as a result of the high temperatures involved. Unfortunately, these oxides contaminate the very metal surfaces being soldered, which interferes with the flow of molten metal and the production of a good quality solder joint.

All electronics-grade solder wire therefore contains an additive called "flux" which helps the molten solder to flow more easily over the joint. It does this by scrubbing away the oxides which arise naturally during heating, and it will often be seen as a pungent brown fluid bubbling away on the joint, accompanied by some fumes.

Those coming into electronics from other industries should note that flux is already contained within "cored" electronics solder and on no account should acid flux be applied separately before using the soldering iron. Plumbers, for example, apply flux paste to copper pipes before soldering them, but electronics-grade solder wire already contains a flux and extra flux is almost never needed. Electronics is no place for acid fluxes!

A close-up of electronics-grade multi-cored solder. Five cores of rosin flux can be seen running through it.

For almost all electronic hand assembly, solder wire containing "Rosin flux" is used. Cores of flux run through the solder wire like letters running through seaside candy, and they prevent the hot area from being contaminated by oxides, otherwise solder would never flow properly and the result would be an incomplete and unreliable joint.

Flux dispenser pens are sold that allows special liquid flux to be applied separately onto a work area. These might be handy for difficult or challenging jobs to help solder to flow better: adding more flux this way won't do any harm and may help a solder joint to be made more quickly and reliably. In my hobby electronics, there's hardly ever been a time when I felt the need to apply extra flux but it's useful for some very tricky or demanding jobs.

For more demanding or precision work, a flux dispenser pen allows additional flux to be applied.

For example I've used specialist Chemtronics flux dispenser pens on tricky, extra-large solder joints involving very thick wires for lead-acid battery connections where I really struggled to make the solder flow properly. You might also use flux dispenser pens in micro-electronic surface-mount work. The extra flux can only help and won't do any damage, but for the rest of the time rosin-flux core solder wire contains sufficient flux and that's all that you'll need.

One thing that many seasoned electronics enthusiasts will recognise is the distinctive smell of rosin flux: its intense woody pine smell is not unpleasant, and flux fumes themselves are not known to be harmful but solder smoke can be an irritant, especially if you suffer from asthma or other respiratory conditions.

Some people can be allergic to rosin: it's used in some sticking plaster (band-aid) adhesive so if they tend to cause skin rashes then you might also be allergic to solder flux. Refer to the First Aid section at the end of the book for some guidance.

Rosin flux is also known as Colophony. It's an amber resin that's glassy and brittle like sugar candy, distilled from the resins of conifers (mainly pine trees) and it's worth knowing about. For electronics use, Colophony is available in small tinlets of solid resin, e.g. LK20 Kolophonium in 20g tins by Donau Elektronik GmbH, from Westfalia or Conrad, or "Kalafonia" is readily available in tinlets on eBay.

Above, Colophony or rosin flux is available in small tinlets.

Solids of Colophony can be dissolved to make a semi-liquid flux, for dipping or applying manually prior to soldering. To make your own rosin flux from solids of Colophony, chip off some fragments and crush them into a small tinfoil dish, then apply some Isopropanol alcohol (start with 2 parts solvent to 1 part Colophony) and let it dissolve over about 20-30 minutes or more.

The following photo sequence shows what happened when I made my own rosin flux this way.

Chip off some solids of colophony and put them in a small tin foil dish.

Above, add e.g. x 2 volume Isopropanol alcohol and allow the solids to dissolve. Experiment as needed.

The resulting flux can be very sticky so handle with care and do not spill the residue.

In this form it can be applied directly with a brush or dropper to areas prior to soldering. In an open dish, after a day or two the solvent will mostly have evaporated to leave an extremely sticky resin that should be handled carefully: it's no co-incidence that Colophony is also used in adhesives and varnishes, and different grades of Colophony resins are used to treat violin or double-bass bows to add friction to bow hairs! Try experimenting with different ratios of solvent and making some up and storing it in an old nail varnish bottle with brush, or generally do what works for you.

Unused Colophony crystals can be reformed into their storage tinlet by warming carefully with a hot air gun, but don't overdo it. You can also use flux cleaners or PCB solvent cleaners if you want to remove traces of excess rosin flux after soldering: these are left behind as a brown deposit and are otherwise harmless.

Temperature Flow

The subject of using your soldering iron to raise the temperature of the materials is discussed next. The aspect of temperature flow takes a bit more understanding, but with practice you'll soon understand how temperature flows within a workpiece being soldered and how you have to harness it properly. After ensuring all parts are clean and the soldering iron is ready to go, the next step to successful soldering requires that the temperature of all the parts is raised to roughly the same level, before solder can be applied.

Imagine the most basic task of soldering a simple resistor onto a printed circuit board: the copper p.c.b. and the resistor lead should both be heated together so that the solder will flow readily over the joint. Later I'll show the precise stages step by step.

A beginner will often mistakenly just heat one part of the joint (e.g. a resistor's wire protruding through a printed circuit board) and hope that the resultant "blob" of solder will be enough to tack everything together. That's completely wrong, because the remaining metal in the joint is cold when molten solder is flooded on to it. The joint will be weak, incomplete or unreliable. Flux will not have flowed properly either, so the joint could be contaminated internally.

Another beginner's mistake is to use a soldering iron to carry blobs of molten solder over to the joint, as if to daub solder over it. The secret of success is to control the iron accurately and apply the hot tip onto the workpiece so that it's in contact with all the parts.

Within a fraction of a second, heat will conduct from the iron and raise the temperature of the entire joint, after which solder can be melted over it. Remove the iron and let the joint cool naturally.

It takes some practice with your chosen soldering iron tip to obtain the best heating action, making sure the tip is clean and tinned properly to begin with.

The melting point of traditional tin-lead solders is about 190°C (374°F) but lead-free tin-based solders require higher temperatures, having melting points of typically 201-227°C (393-440°F). As Antex reminds us, the melting point is not the temperature of the soldering iron tip: instead you should set a temperature that ensures the solder melts instantly onto the tip.

Fixed-temperature soldering irons have no adjustability but they'll cope just fine with either type of solder. A soldering station usually has a variable control that gives more control in different circumstances. In practice the iron tip temperature should be set for typically 330-350°C (626-662°F) or maybe a little more if using lead-free solder.

Set a typical temperature of 330-350 °C (626-662 °F) or more, if you have a soldering station.

The next diagram shows what would happen if you applied a hot soldering iron to an imaginary metal block. **Heat travels from hot to cold.** In step (1) heat travels out of the iron into the cold metal block, which starts to warm up starting at the edges. Gradually (2), the whole metal block's temperature rises until the middle of the block finally rises in temperature as well.

In effect heat is now travelling back "towards" the iron tip, until finally (3) the whole block is at the same temperature as the iron. At this point solder could now be applied.

You'll notice this effect as a time delay when soldering any joint. The "working area" of the iron's tip gets cooled to begin with, because the metal in the rest of the joint is sinking heat away from where it's most wanted.

Only after the whole joint has reached the melting point temperature of the solder can solder be melted onto it effectively.

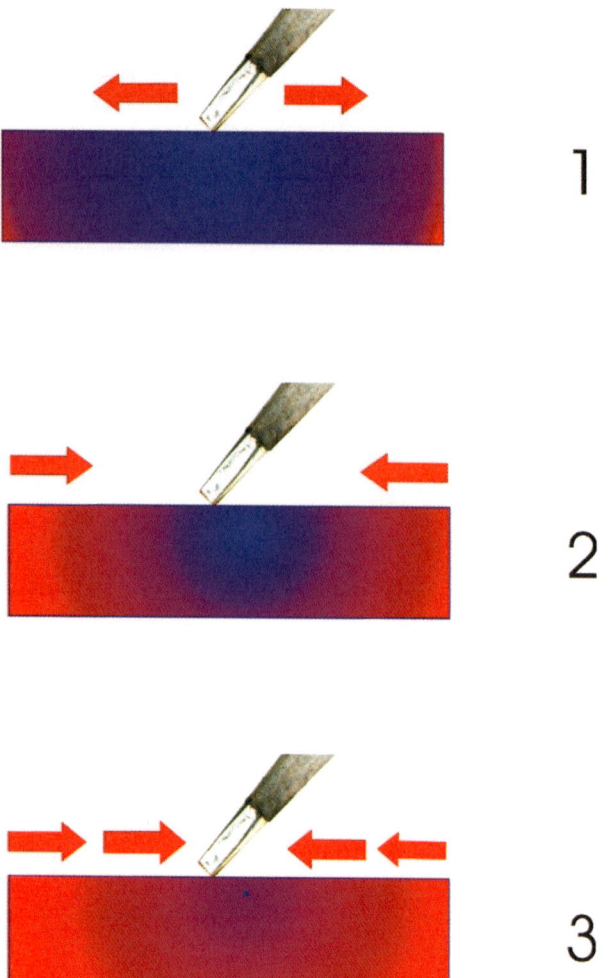

How metal in the joint actually "sinks" heat away from the tip to begin with. In (1) the soldering iron is applied to some cold metal which sucks (or 'sinks') heat out of the iron. In (2) the furthest-away parts of the cold metal start to heat up.

As the metal continues to warm up, in effect heat moves back towards the tip until the whole piece has reached the solder's melting point (3) when solder can be applied successfully.

Soldering a small p.c.b. joint

With experience, you'll get a feel for how long it takes after applying the iron before you can apply solder. The more metal that is present in the joint, the longer the period that heat must be applied. A small p.c.b. joint takes well under one second to complete. A large metal terminal could take quite a few seconds or more to heat up.

As I explained in the introduction, higher power (wattage) irons cope better with larger workpieces because they recover more quickly and are more "unstoppable", making it easier for them to heat larger workpieces without cooling down so much.

If you apply solder too early, it won't melt properly and the result will be a grey, crystalline joint caused by the solder's melting point temperature not being reached and the flux not having flowed properly. Semiconductors must be soldered as rapidly as possible as they are heat-sensitive, but they're a lot more robust than they used to be.

Until they have gained some practice, novices sometimes buy a small clip-on heat-shunt (next photo) which resembles a pair of aluminium tweezers. In the example of, say, a transistor, the shunt is attached to one of the leads near to the transistor's body. Any excess heat then diverts up the heat shunt instead of into the transistor junction, thereby avoiding the risk of thermal damage. Applying far too much heat may destroy the part or damage the p.c.b. foil which may lift away from the board.

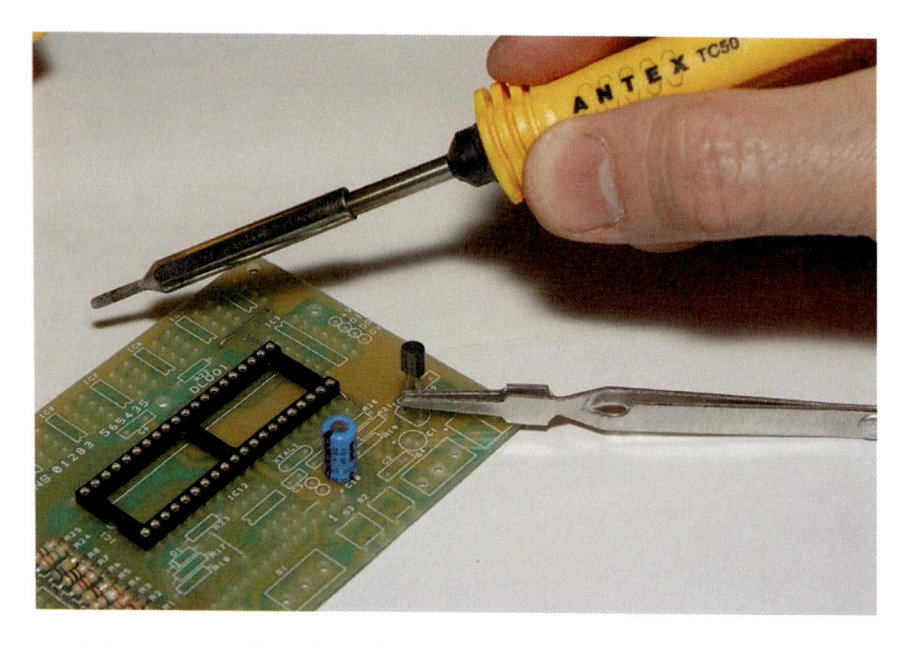

A clip-on Antex heat shunt fitted to a transistor leg, helps prevent thermal damage due to overheating when soldering it in place. Less essential these days, but beginners find them re-assuring.

In due course constructors also learn to judge how much solder should be applied to any particular joint. An ideal p.c.b. joint is very slightly concave in shape. If not enough solder is used, the result may be an incomplete joint which may cause an intermittent fault later on. An excess of solder – shaped like a ball bearing - is an unnecessary waste and in extreme cases may cause short circuits, especially on densely-populated boards. There is no need to add more solder "for luck".

Professionally-produced p.c.b.s have a green solder resist coating which helps to ensure that solder does not stray onto adjacent pads. As a finishing touch, I usually spray the solder-side of a circuitboard with aerosol spray lacquer afterwards. It keeps the solder joints nice and shiny and helps prevent corrosion.

It is worth remembering that some electronic components can create hazards of their own during the soldering operation. For example, several times I have soldered what turned out to be an energized, or charged, component such as a large capacitor and the molten solder has completed the circuit to the device. This caused tiny splashes of molten solder to be ejected due to arcing, risking eyesight damage. Such events are extremely rare but it's worth knowing about and some typical hazards are listed on the following page.

Component Hazards

Some components can create hazards during a soldering operation:

Coin cells and **button batteries** are commonly used as power, real-time clock or memory backups. If heated excessively they can explode without warning due to the build-up of internal pressure. Spot-welders are used in industry to connect tags to them, but if you need to solder wires to such a cell (especially small ones) then it should be done as quickly as possible. You would have to be extremely unlucky or very careless to experience a problem but it's best to use p.c.b.-mounted battery holders instead.

Some **memory back-up capacitors** or **electrolytic capacitors** remain energised for a while even after the circuit has been powered down. Be especially wary of any capacitors that have high voltage ratings (e.g. 400V or 600V) marked on them as this alludes to high voltages stored within. Molten solder is a perfect electrical conductor and in some cases the component's contacts could be shorted during the soldering (or desoldering) operation. If molten solder shorts it out then the arcing may cause globules of solder to be spattered outwards without warning, potentially risking eyesight damage.

Always ensure that powered components are electrically inert and discharged before soldering. Cells, batteries and battery packs should not be accidentally shorted during the soldering process, to avoid arcing and solder spatter. Note that electrolytic capacitors can also explode after a while if reverse-connected, so observe polarity closely.

In the next chapter the practical stages of soldering components and wires successfully are explained.

5 SOLDERING STEP BY STEP

Earlier I explained that the individual factors that affect the quality of a solder joint are:

- **Cleanliness** – dirt or impurities hinder good solder coverage.
- **Temperature** – the right level to enable the solder to flow freely!
- **Time** – apply heat for just the right amount of time!
- **Adequate solder coverage** – enough to form a good joint without touching neighbouring areas.

These rules apply whether soldering a p.c.b. or performing other tasks such as interwiring (hooking everything together with connecting wire). We'll now summarise the stages of making a typical solder joint – soldering components onto a printed circuit board (through-hole soldering).

Most people insert components into the circuit board and simply splay the wires out to hold them in place under spring tension. I find it best to snip excess wire leads off at this stage, to improve accessibility.

1. All parts must be bright, clean and free from dirt and grease.
2. Try to secure the work firmly to stop parts moving around.
3. "Tin" the soldering iron tip with a small amount of solder. Do this immediately, with new tips being used for the first time.
4. Wipe the tip of the hot soldering iron on a damp cellulose sponge to remove excess solder or contamination.
5. Many people then add a tiny amount of fresh solder to the cleansed tip just before using it.
6. Heat all parts of the joint with the iron typically for under a second or so, until it's heated throughout.
7. While heating, then apply sufficient rosin-core solder to form an adequately-covered joint. (*contd.*)

The Basic Soldering Guide Handbook

8. It only takes a second or two at most, to solder the average p.c.b. joint this way.
9. Do not move parts until the solder has cooled.
10. Remove and return the iron safely to its stand.

The next photo sequence illustrates these stages. It's best to start with the smallest, fiddliest parts first when soldering a blank p.c.b., because that's when you've got the most access on the board. Accessibility will be reduced as more components are added, so we'll start with a simple wire link on a professionally designed p.c.b.

A typical professionally-designed single-sided blank p.c.b. – white silk-screen printing shows what components go where.

The copper track underneath shows through as tan-coloured tracks and larger "lands". The underside has also been treated with a green solder resist coating that confines molten solder to the intended area. The solder pads are ready-tinned to help with soldering. (Courtesy Magenta Electronics Ltd.)

Above, preparing a wire link for soldering – cut off some tinned copper wire and bend the ends to fit the p.c.b. correctly. Round-nose pliers are ideal for this, but electronics or "radio" pliers are fine.

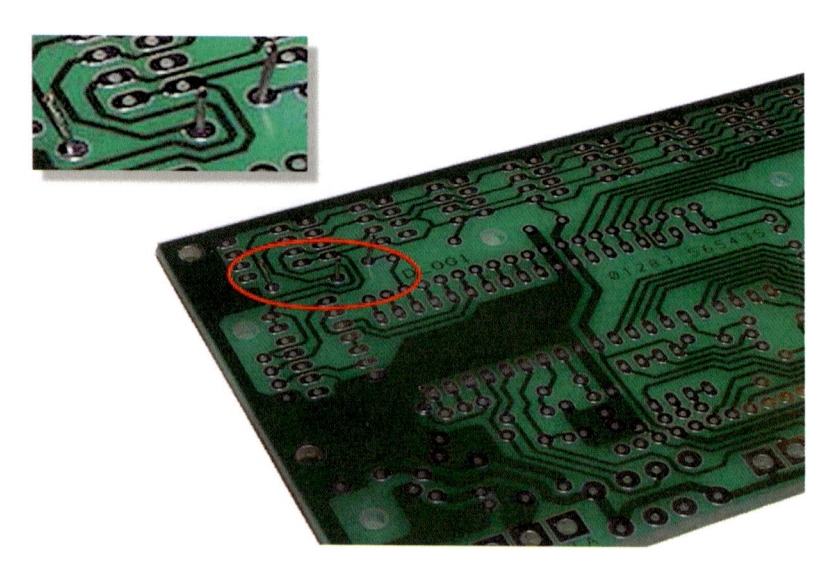

Wires are prodded through the holes (circled) in the board, then turn it upside down to view the solder side. Splay the wire ends apart slightly, so they are held in place while you solder them.

The Basic Soldering Guide Handbook

Above, wipe the hot soldering iron on the damp sponge to clean the tip. Do this periodically when contamination, flux deposits etc. build up on the iron to keep the tip nice and shiny.

Dabbing a tiny amount of solder onto the tip improves heat transfer. The molten solder fills the void between the hot iron and workpiece.

Then apply the soldering iron to heat both the solder "pad" and the wire end at the same time (say < 1 second). Apply a few millimetres only, of solder. Then remove the soldering iron immediately and allow the joint to cool down naturally.

Tidy up the component wiring using wire-cutters, sometimes called "side-cutters" or "end-cutters" depending on the design of the cutter's blades. End-cutters (see next photo) are a bit of a luxury item and aren't essential, but they help to create a neat, flush cut.

Note! Wire off-cuts can 'ping' off at a random angle after snipping and they are a notorious risk, so take extra care to ensure that they do not cause an eye injury.

The Basic Soldering Guide Handbook

These "end cutters" can then be used to snip off excess wire...

... but ordinary electronics "side cutters" or wire cutters are fine too.

Alan Winstanley

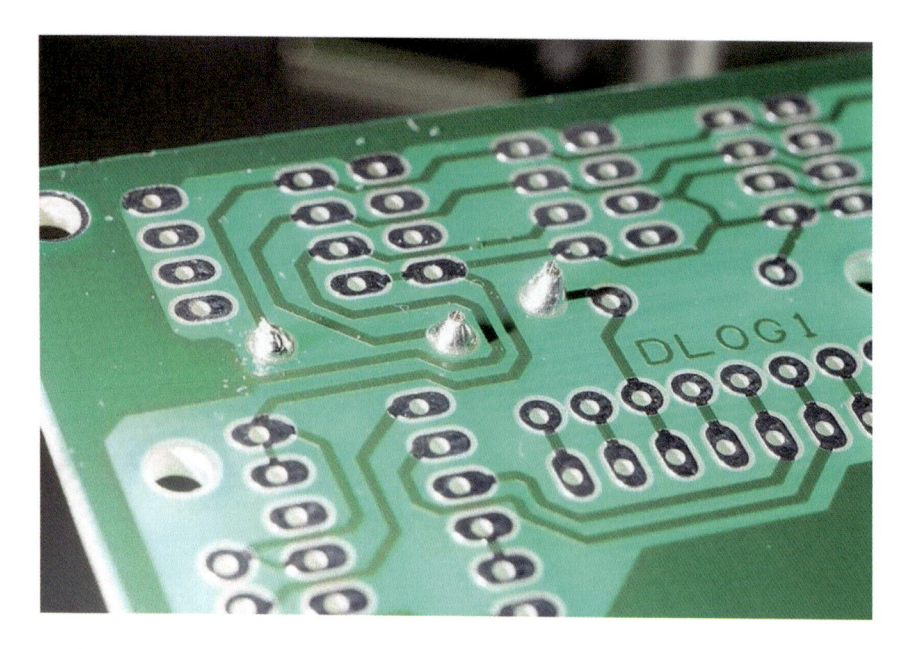

The ideal solder joint should be smooth and slightly concave, not "ball-bearing" shaped and quite shiny, not dull or crystalline-looking.

A large i.c. socket can be added next...

The Basic Soldering Guide Handbook

Ensure the i.c. socket is flush against the board before soldering. You can "tack" it into place by first soldering the pins located on diagonally opposite corners of the socket, to hold it flat on the board.

As hardly any metal is involved, each pin should take well under a second to solder at most.

Try to get consistent results in your soldering, but don't worry too much about the lack of uniformity – this is soldering by hand after all, not by precision machinery, and some slight variation is OK provided the solder coverage is good and the soldering is generally clean and effective. You'll improve with practice, that's for sure!

Continue until the i.c. socket is soldered into place.

Once the smallest parts are soldered into place, you can continue to solder the remaining components. It's easiest to handle the smallest so-called "discrete" parts first while you still have plenty of room on the board. I usually solder resistors and capacitors next. The principle of soldering them is just the same as a simple wire link: insert them from above till they are flush on the board, then splay their wires a little to hold them in place, and preferably snip off at least some of the wire to give you more access.

The next photo sequence shows more components being soldered onto the printed circuit board. As more parts continue to be added, accessibility to the board will become more restricted, so it's easiest to solder the smallest parts first.

The Basic Soldering Guide Handbook

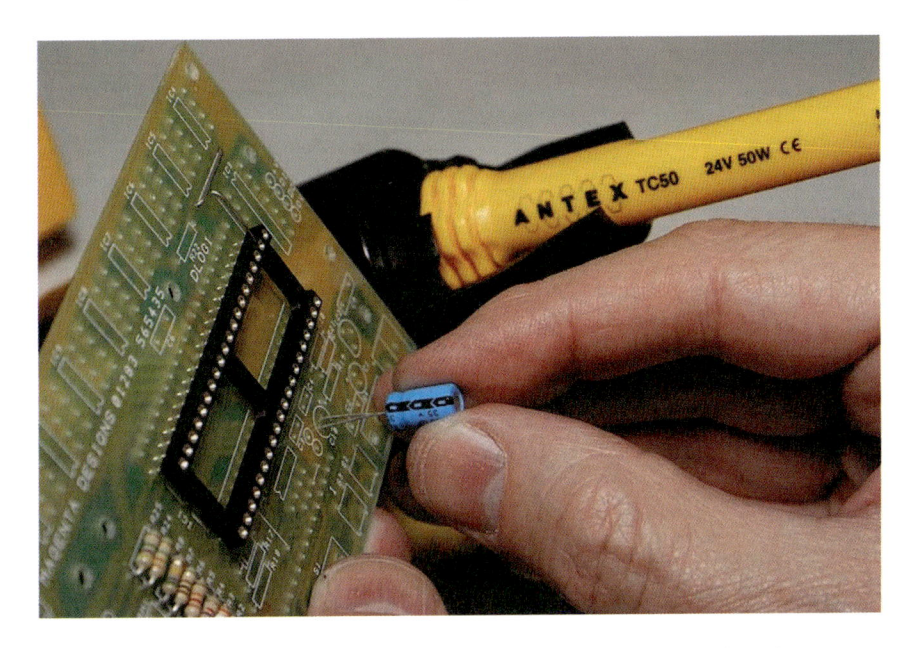

Continue by inserting "discrete" components from above, splay their wires out underneath to secure them in place, then snip and solder the joints exactly as before. This blue electrolytic capacitor is polarity sensitive (note the – sign to denote the negative terminal).

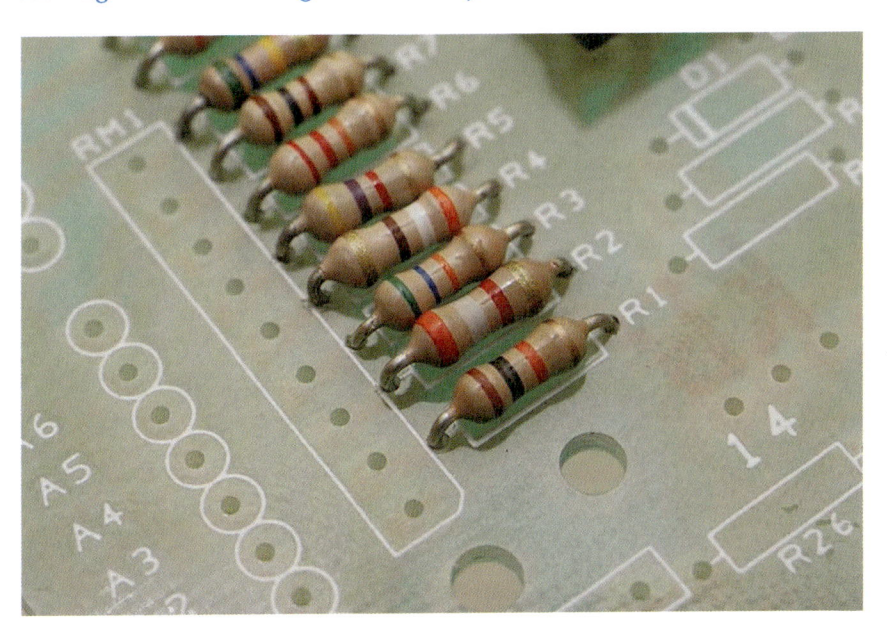

This row of ¼ watt resistors was soldered into position next.

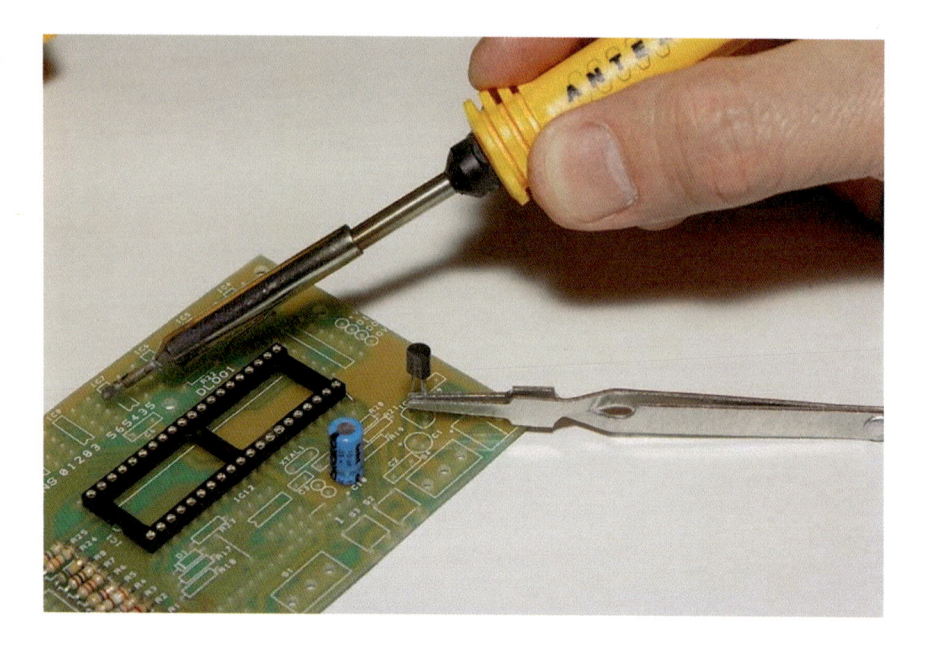

This transistor was next. Solder it quickly to avoid thermal damage, and orientate it correctly. The clip-on heatshunt is optional and it diverts heat away from the temperature-sensitive semiconductor.

Unusually, this toggle switch fitted directly onto the board as well.

As there is comparatively more metal to heat up, it'll take longer to solder the switch terminals, and you'll need more solder as well. Thicker gauge solder is useful at such times. Allow say 2-3 seconds to solder each terminal.

Don't forget to clean the soldering iron tip on its damp sponge every now and then, to ensure the bit is kept clean and shiny.

With practice, through-hole soldering of p.c.b.s will become second nature. There's no substitute for tackling some soldering jobs though, particularly trying a simple kit based on a quality p.c.b. which will boost your confidence enormously. Later on I'll show you how to correct any problems by desoldering using various techniques. Next, we'll move on from "through-hole" soldering techniques and look at how to handle wires and leads successfully.

Most electronic devices need connecting up to external components such as battery packs, speakers, l.e.d.s or switches. Usually, multi-stranded connecting wire is used to connect circuit boards and external parts together. Unlike single solid-core wire, multi-stranded wire is flexible and vibration-resistant. Hobbyists mainly use 7/0.2mm wire (7 strands, each 0.2mm diameter) or similar for low-voltage hook-ups although much Chinese equipment uses much thinner wires than this.

In the following photo sequences I show how a potentiometer (a panel-mounted variable resistor) and a light-emitting diode (l.e.d.) are connected using multi-stranded wire. The same principles of soldering apply to most other components including panel-mounted switches, loudspeakers, buzzers, audio sockets and more.

Components usually have terminals or "tags" onto which wires can be soldered. Start by ensuring the component's tags are clean otherwise solder will not wet properly, and the joint will be impossible to solder. This is especially true of parts that have been in storage for a long time. The connections often oxidise or blacken, so clean the solder tags with e.g. an abrasive glass-fibre brush (see Chapter 3), or a needle file or abrasive paper.

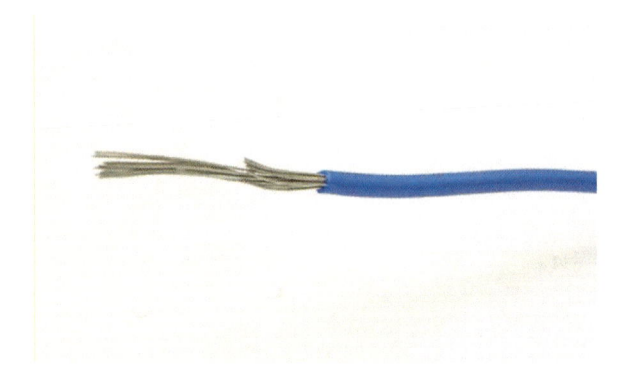

How not to strip insulation off wire: some of the copper cores have also been cut – avoid doing this!

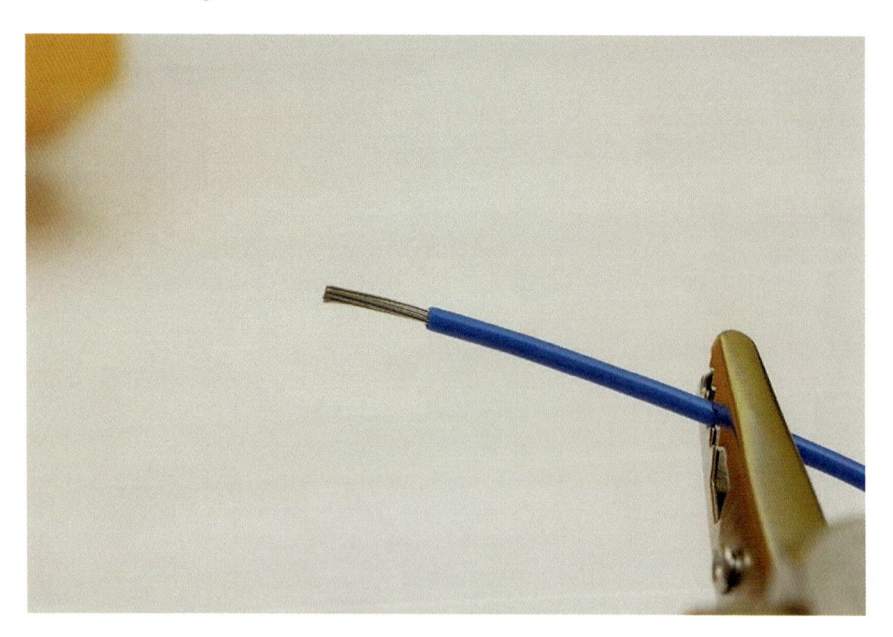

A "Helping Hand" croc clip can help with soldering.

Start by removing a short length of insulation from the connecting wire using wirestrippers. There are then two ways to solder it to a component's solder tag. The first way is to tin the stripped wire end to solidify it - just heat it with the iron (next photo) and dab a little solder on it, and let it cool. Poke the resulting wire end through the solder tag, heat both parts with the iron and solder them together using a few millimetres of solder.

Although the assembly doesn't hold itself together so well during soldering (consider a Helping Hands jig if needed) this is quicker and easier to make and also easy to desolder again, and is perfectly adequate for most joints of this kind. The majority of commercial wire joints seem to be made this way.

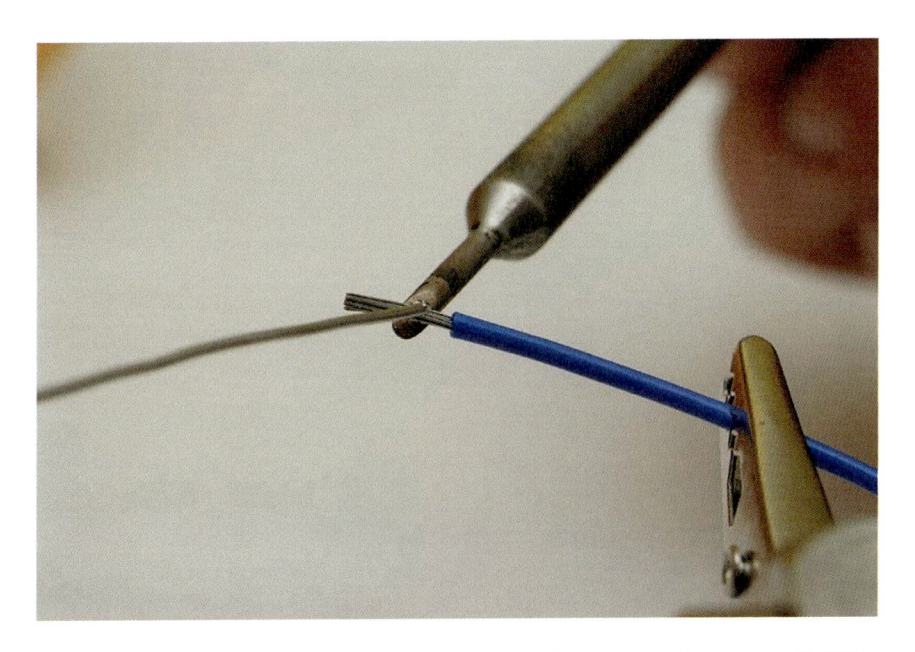

Apply a hot iron and then some solder, in order to "tin" the wire ends. This makes them into a solid. Applying the iron for too long will cause the wire's PVC insulation to melt and shrink back, so try not to heat the joint for longer than you need to.

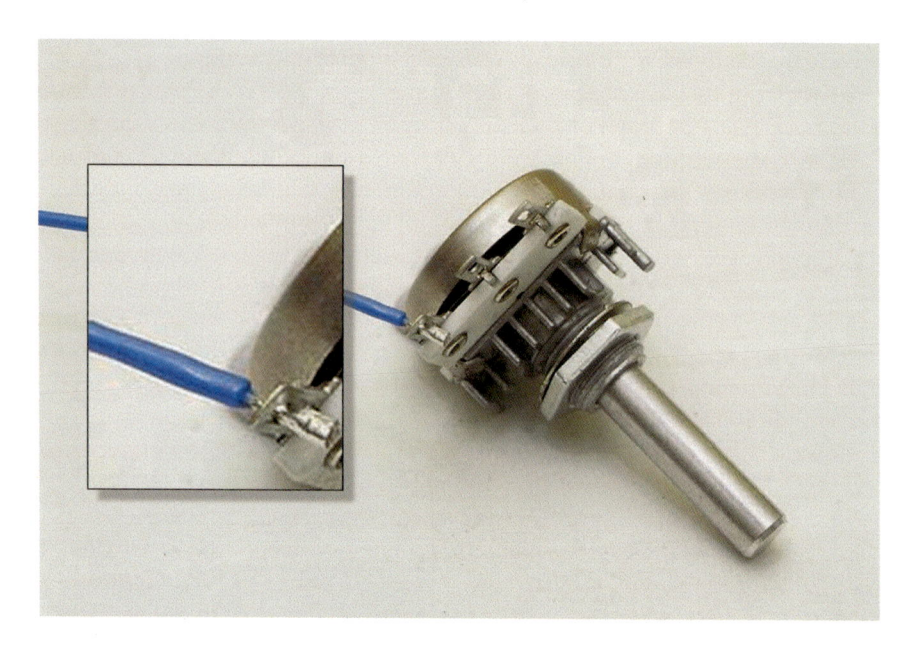

Then poke the tinned wire through the hole in the solder tag. Shorten the tinned wire end with wirecutters if needed...

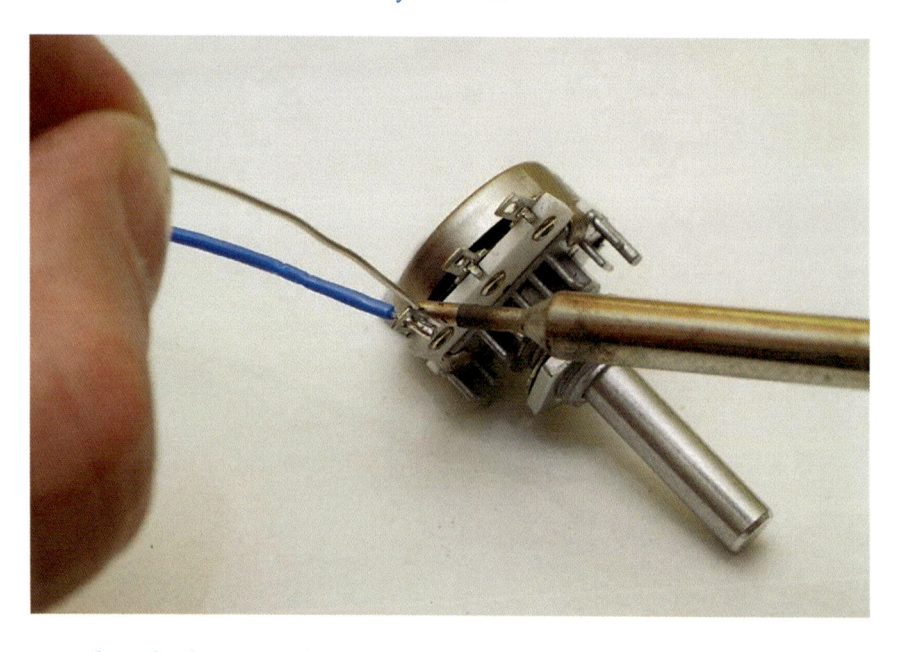

... and apply the iron to heat the joint. As there's quite a lot of metal to heat here, allow 2-3 seconds to heat it first and then simply solder the wire and solder tag together with a dab of solder wire.

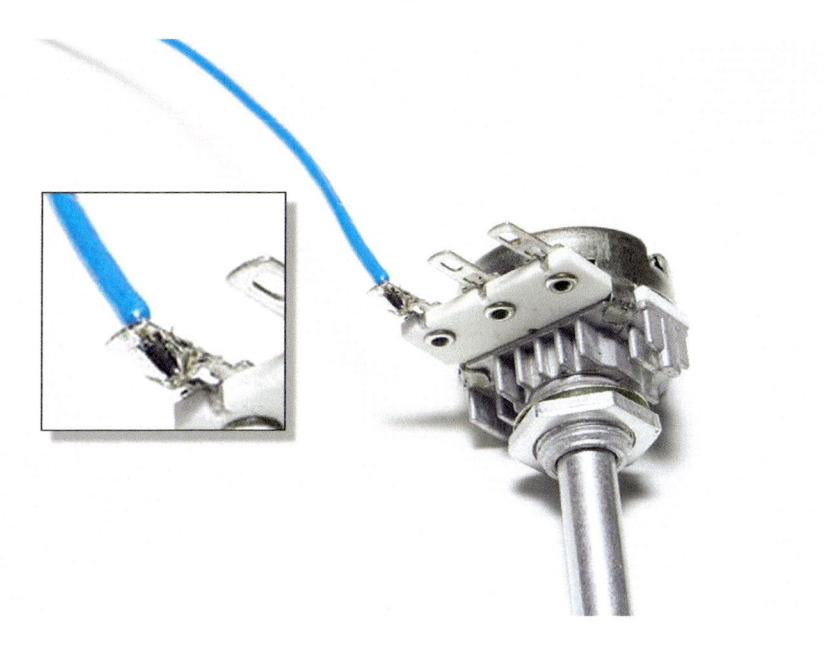

The result is a perfectly satisfactory solder joint.

The second way is to loop the **untinned** wire through the tag first. This secures the wire during soldering, but desoldering it again is trickier:

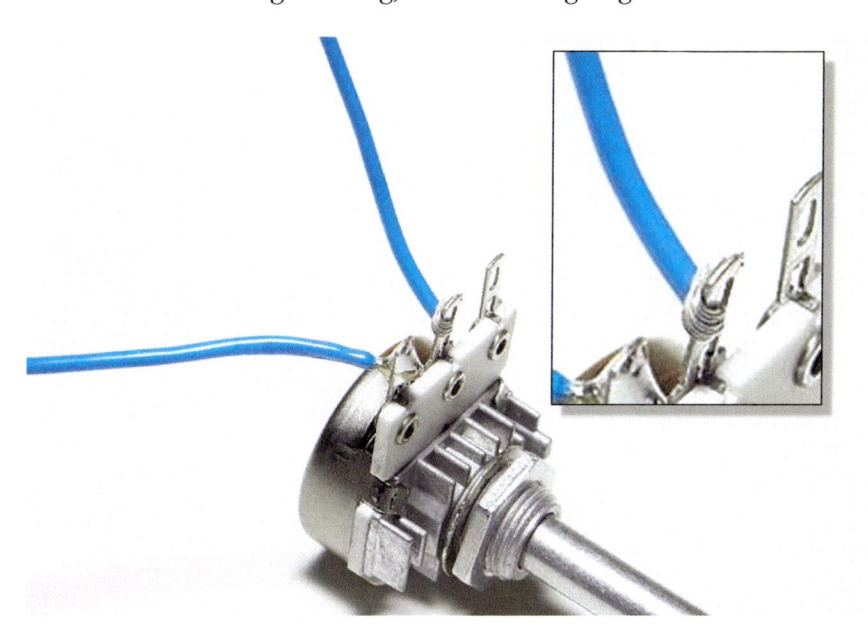

This wire is wrapped securely around the tag before soldering.

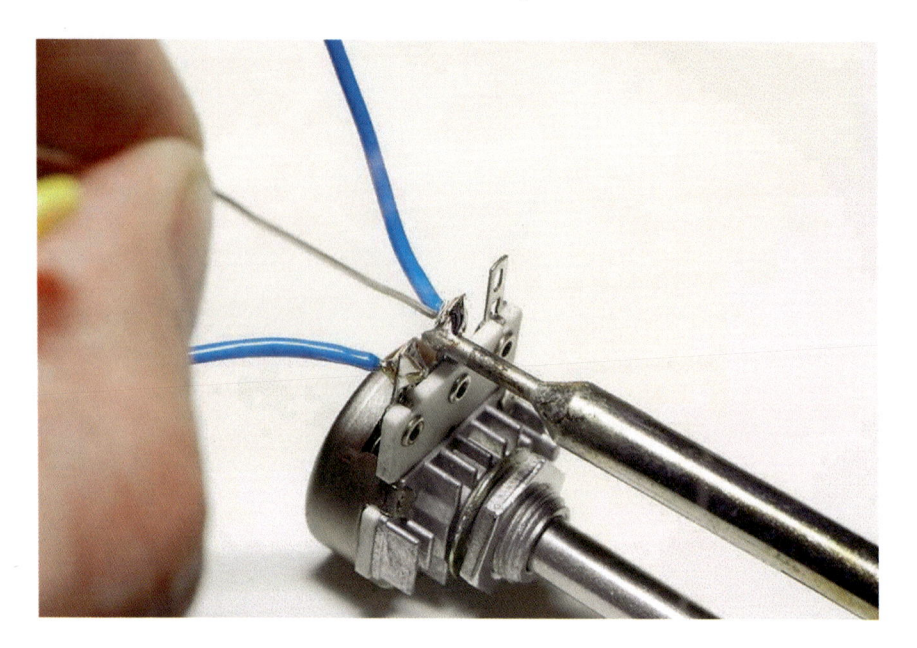

Solder the joint in just the same way – heat the tag and wire simultaneously and dab some solder wire over them.

Much electronic equipment is connected or inter-wired this way and soldering everything together is a key stage in assembling any electronic project. Users can decide whether it's best to pre-tin a wire before soldering it to a tag, or wrap bare wires around terminals or tags first before soldering everything together.

Soldering isn't always the best solution to some problems, for reasons that I explain in Chapter 6 in a section called *Fatigue and Breakage*. However you can solder several wires together to join them simply and cheaply, and next I show a classic wire joint of this kind, sometimes called a Western Union joint (next photo).

Strip the ends from each wire using wire strippers, and avoid cutting any of the stranded cores or this will create a weakness. (Stray strands of wire can also be a hazard, especially at high voltages.) Then twist the wires securely together, ideally four or five times. Apply a soldering iron to the exposed joint and dab on some solder until it flows fully over the joint, then remove the iron and allow the joint to cool naturally.

The wire's plastic insulation may shrink back a little due to the heat, but try not to overheat it excessively. If you have the resources you can finish off by insulating the joint with some heatshrink tubing (slide it over a wire before you make the solder joint!). Otherwise use ordinary PVC tubing or insulation tape if you have any.

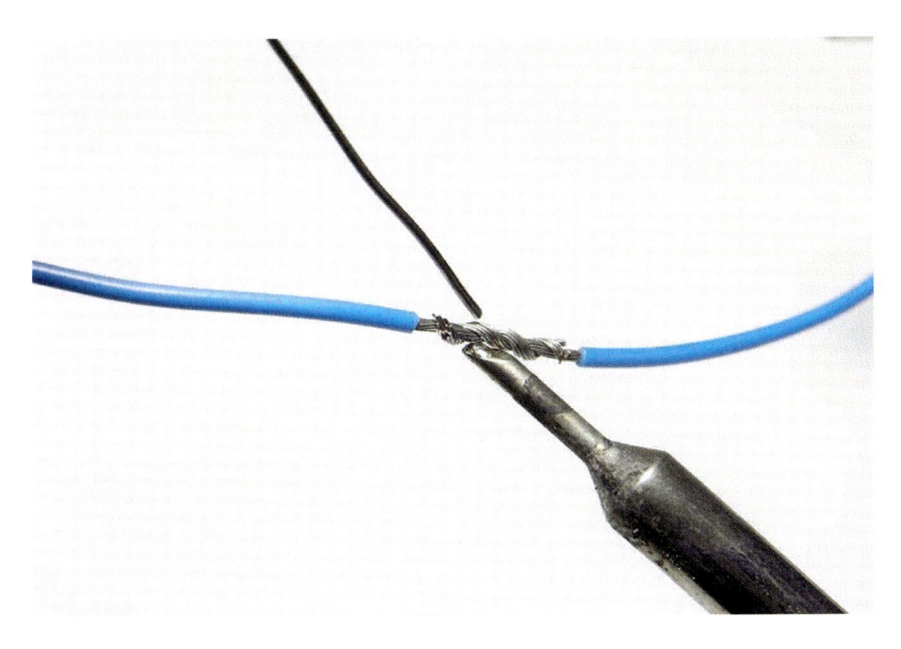

A Western Union joint is a basic way of soldering wires together.

Solder tags are used to connect wires to e.g. a metal panel. Twist a stripped wire over several turns so it holds itself together when soldering.

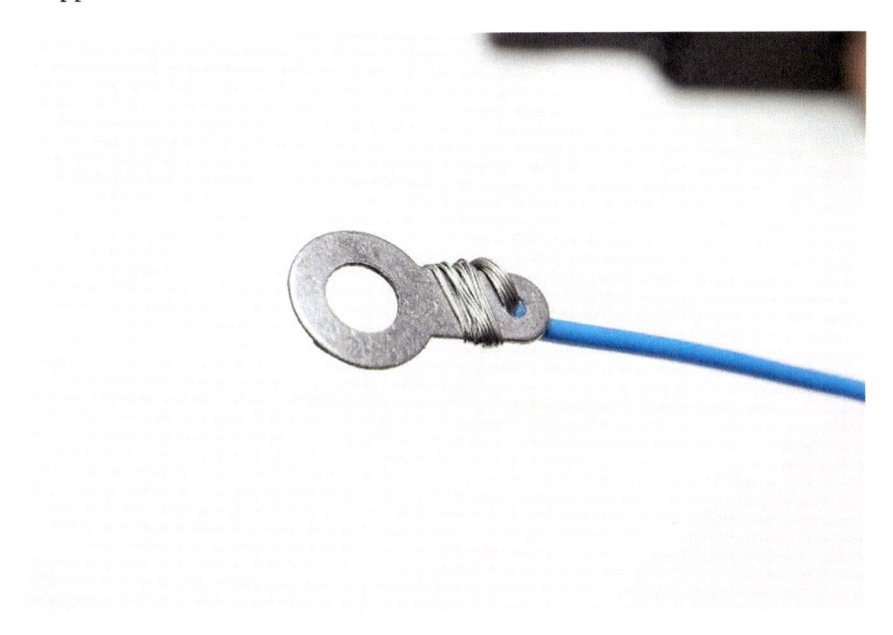

A stripped wire is wrapped several times over a solder tag to secure it.

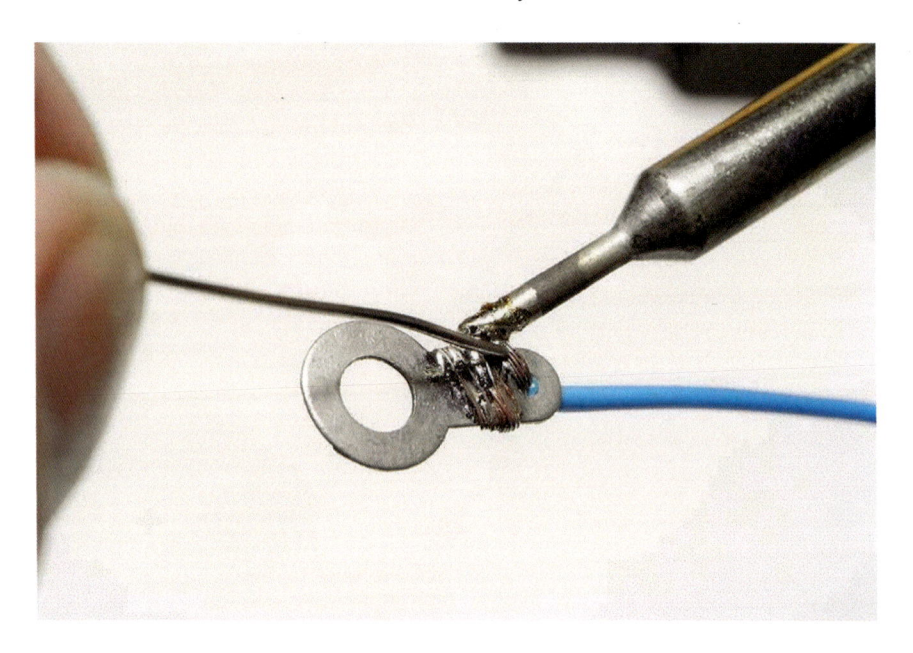

Apply plenty of heat and solder for a few seconds. Expect the PVC wire insulation to shrink back a little due to the heat, but try to avoid this as far as possible.

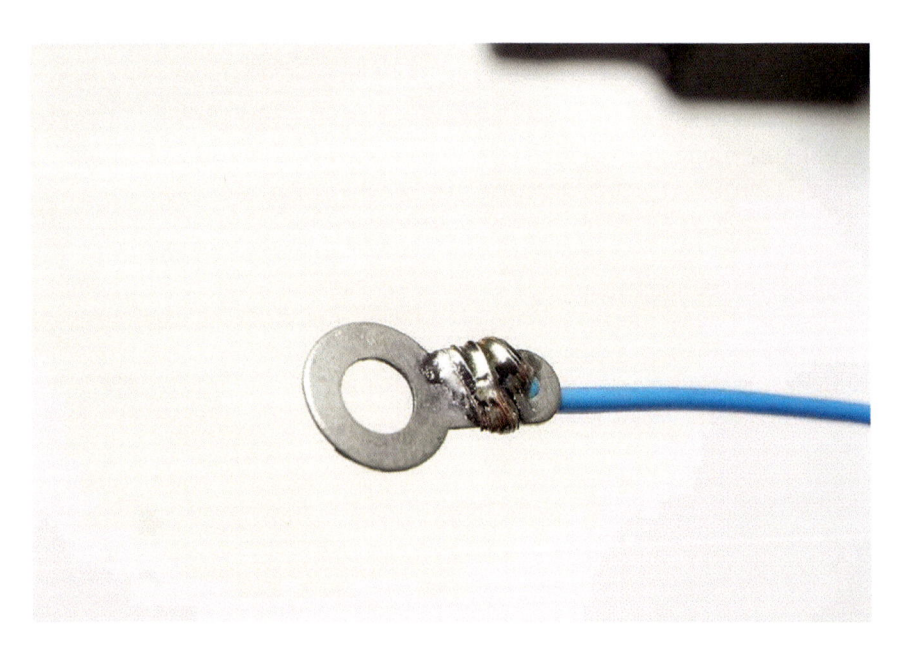

A completed solder tag. Not recommended if vibration is an issue.

The next sequence show how the same basic principles of soldering wires are used to connect up a device such as a panel-mounting l.e.d. A current-limiting resistor, if used, can be wrapped around and soldered directly to the anode lead, and then two multi-stranded connecting leads can be soldered direct to the l.e.d. cathode and resistor. Use coloured heatshrink tubing or PVC sleeve to prevent short circuits.

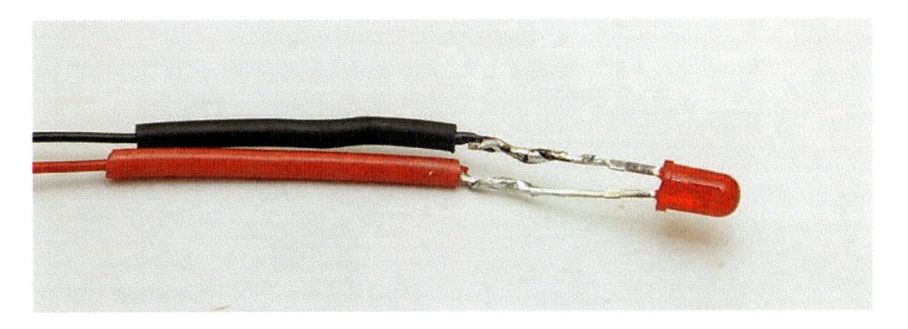

Multi-stranded wires can be wrapped around and soldered directly to the solid leads of an l.e.d. Trim off any excess as needed.

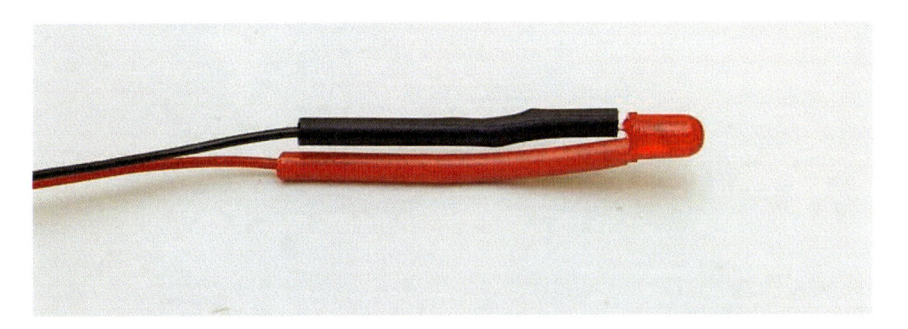

PVC sleeving or heatshrink is needed to prevent short circuits.

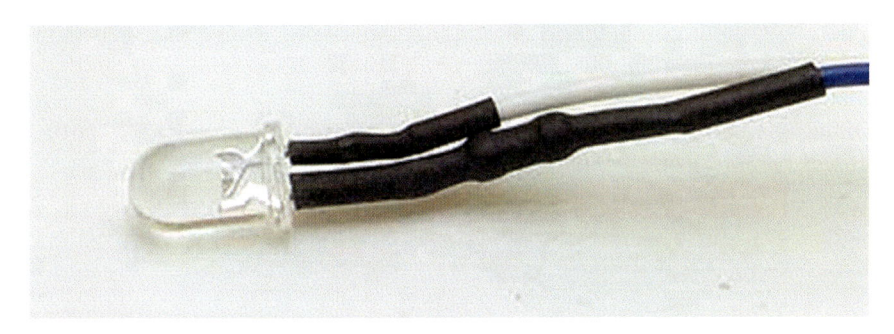

A commercial l.e.d.- the series resistor is visible in the sleeving.

Wires can be through-hole soldered direct to printed circuit boards by stripping and tinning the ends, and this is a very common and cheap way of hooking up a board using "flying leads" which you'll see this all the time in consumer electronic equipment.

Obviously the tinned wire has to fit through the hole in the p.c.b., and more than once I've struggled with nuisance wires that are just too wide to fit after I've tinned them: the solder has made them too large to pass through the hole. Sometimes I crush the tinned end with pliers a little to make them fit through, or otherwise I have to start again.

So-called "solder pins" can also be used, onto which flying leads may be soldered from the *component-side* instead. Solder pins can be more convenient during assembly as it saves having to constantly turn the board over to solder wires to the underside.

A set of connecting wires for an l.e.d., stripped, tinned and ready to be soldered to the p.c.b.

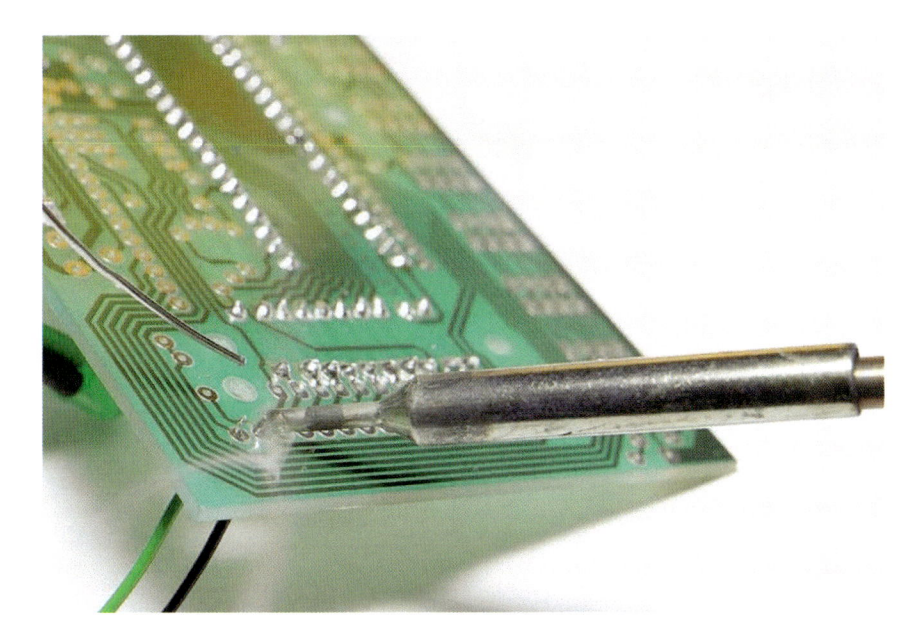

Solder the two "flying leads" to the copper pads in the usual way. Hopefully the holes are big enough for the wires! If not, all you can do is enlarge the hole if possible, or reduce the thickness of the soldered wire somehow to make it fit through.

Wires can also be attached to the *underside* (solder-side) of circuit boards, simply by tinning and tacking the ends onto existing solder joints, by re-melting them and absorbing the tinned wire end into it ("reflow soldering" – see later). This is a cheap and cheerful, semi-reliable way of rigging wires to a circuit board, and it's used all the time in imported consumer electronics.

For example, negative supply wires might simply be tacked onto the p.c.b. using the solder joint of a component that's connected to the negative rail. The solder joint is re-melted and the wire is just tacked onto it.

It is usually best to cut off excess wire after soldering. I prefer to tidy up the joint by snipping any excess wire from the joint using a pair of end cutters shown earlier. These expensive hand tools have specially angled blades that snip the joint flush to the top of the solder joint. Ordinary cutters will do just fine though.

Using a through-hole soldering technique to connect flying leads to a printed circuit board.

It's worth taking time out to inspect the work closely, looking for any missing solder joints, whiskers of solder or swarf shorting out any solder pads, and all such potential problem areas should be dealt with prior to testing the board.

Faultfinding goes beyond the scope of this guide, but it's true to say that almost always, any problems noticed after powering up the circuit are due to soldering faults or wrong components being used or connected wrongly.

Hopefully readers have now gained an understanding of the basic steps of soldering electronic components and hook-up wires. There is however no substitute for practice and with experience readers will soon gain the necessary confidence to solder successfully.

6 REFLOW & DESOLDERING TECHNIQUES

Another technique often used in electronics assembly is "reflow" soldering. This is used to "tack" devices or wires together, especially if very small, sensitive or fiddly parts are involved; there might be no room to make a "proper" sturdy joint, or it might not need any mechanical strength due to the small sizes involved.

Reflow soldering is also a key part of using surface-mount devices (SMDs), tiny chip-like components that have next to no mass but need to be soldered onto a p.c.b. with great accuracy. SMD assembly is very fiddly and needs good dexterity and soldering skills so it is an area probably best avoided by absolute beginners. It goes beyond the scope of this handbook to describe SMD methods in depth, but the leap from manual soldering described here to SMD work is not actually a great one: surface-mount work needs much more care and precision to ensure success, especially when the work is being performed manually rather than with precision production machinery.

As an example of a simple reflow technique, imagine a small temperature sensor (I used a transistor) for use in a thermometer project. It could be quite tricky to solder flying leads onto the sensor's leadouts, so a good approach is to tin both the flying leads and the sensor's leads, and then simply touch them together and re-melt the solder with the iron. There's no need to add any more solder, because the solder that's already there will melt and the joint will be made. Sometimes this is called a butt joint.

The next photo sequence shows how this manual reflow soldering task worked in practice.

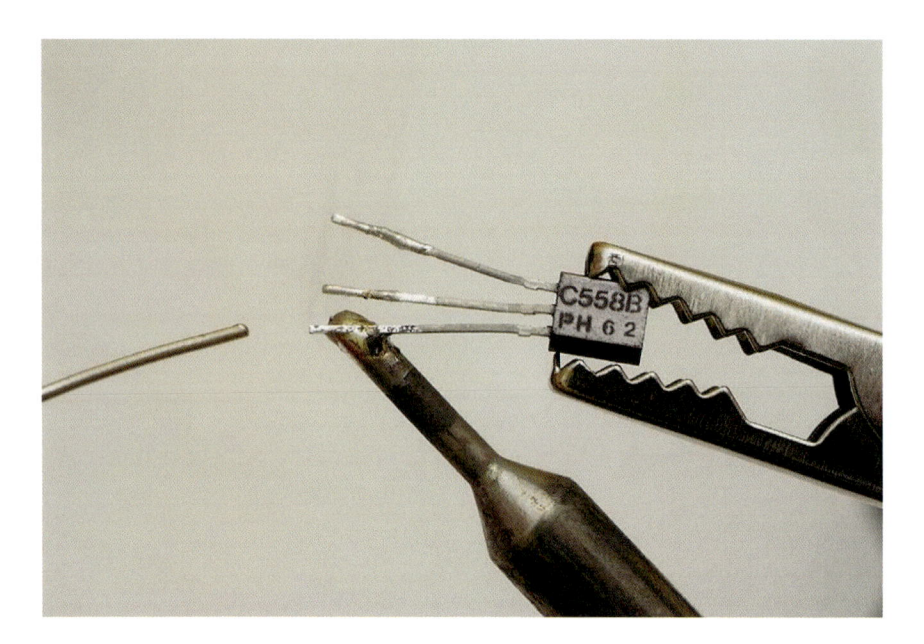

Above, the leads of this sensor (transistor) are being tinned ready for flying leads to be tacked onto them using a reflow soldering method. A Helping Hands jig might help to grip small parts.

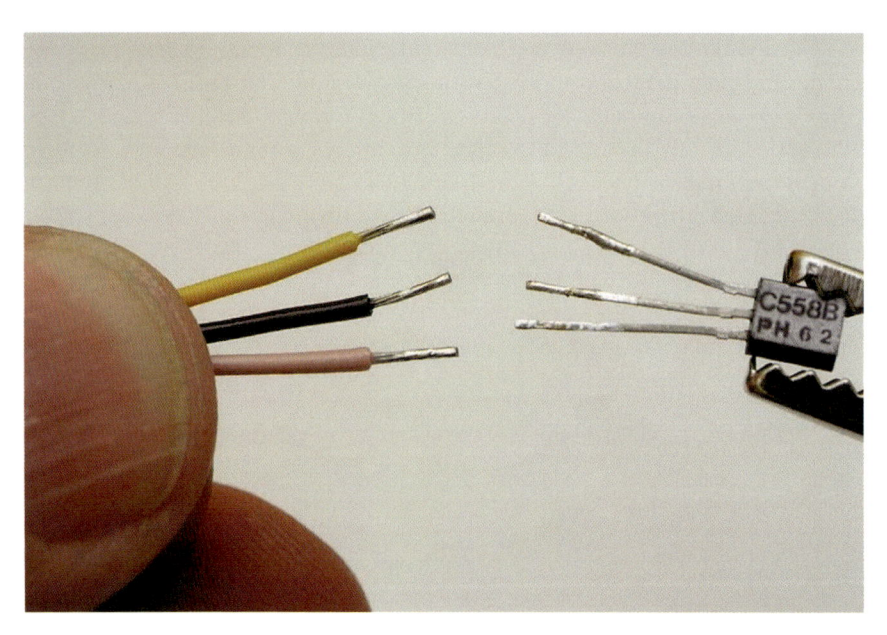

The three flying leads stripped, tinned and ready to be reflow-soldered onto the device, also tinned.

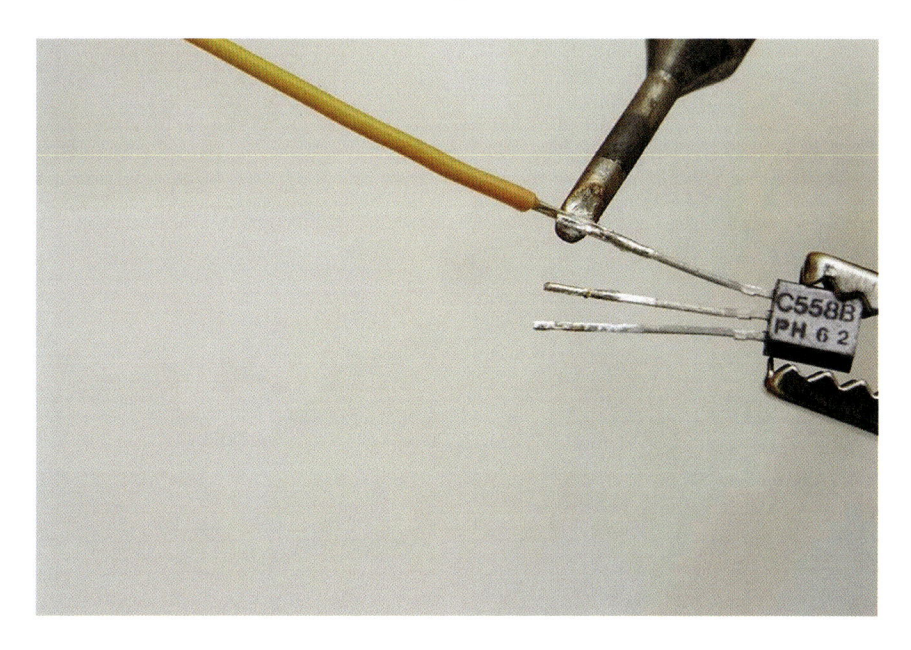

To reflow solder them, carefully hold the two wires together while re-melting the solder with the iron. The solder melts together into one joint.

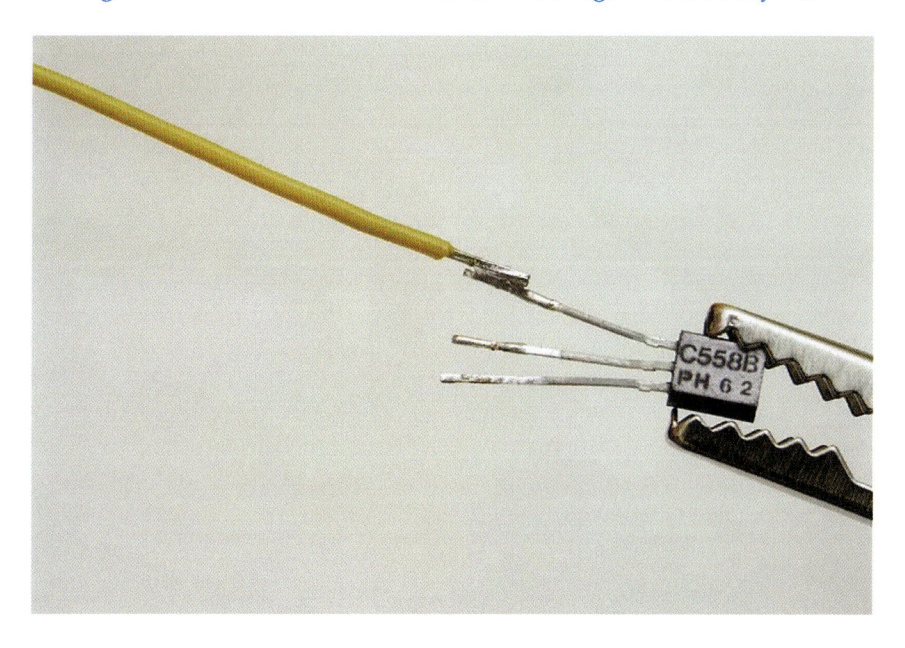

Remove the iron and let the solder cool. The wire is tacked on.

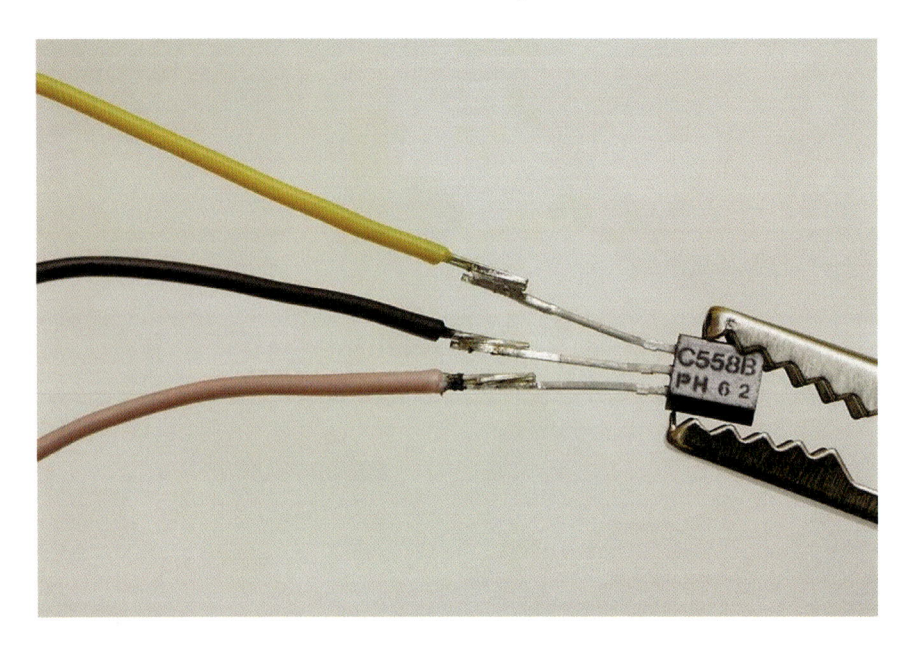

Repeat for the other leads. They just need insulating with sleeving, then the reflow soldering job is done!

Reflow soldering can be very useful for quickly tacking wires or small components together, especially where little weight or stress is involved.

Depicted next is a typical "tag strip", an insulated panel with metal solder tags used for making sundry connections. (Entire TVs and radios used to be hand-built with them, in the early-mid 20th Century!) The principles of soldering are exactly the same, but because more metal is present, more heat is needed to reach the melting point of the solder.

You'll also need more solder for bigger joints like these, so larger diameter solder makes a quicker job of it. Consider adding more flux (see earlier) to see if it helps the solder to flow more easily.

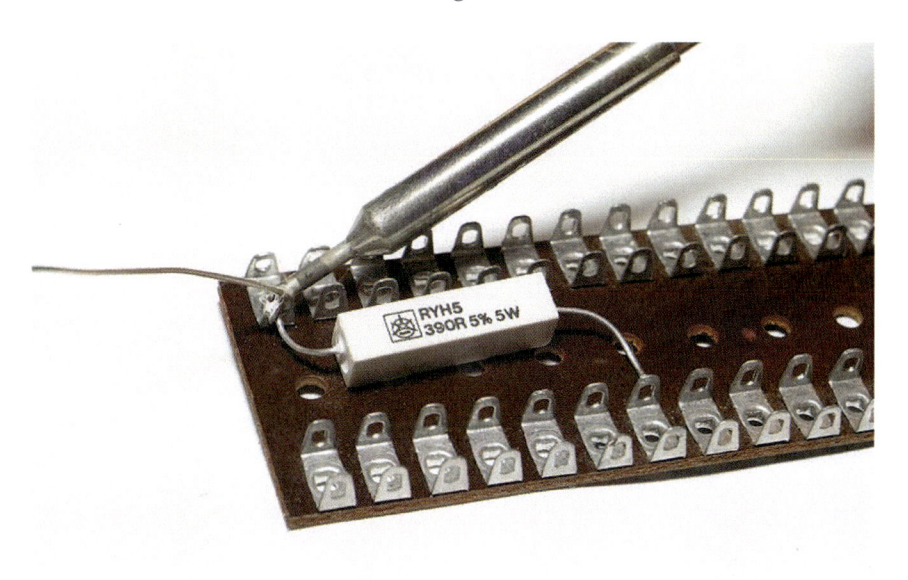

Wrap wires through the terminals and arrange everything neatly, then solder as normal.

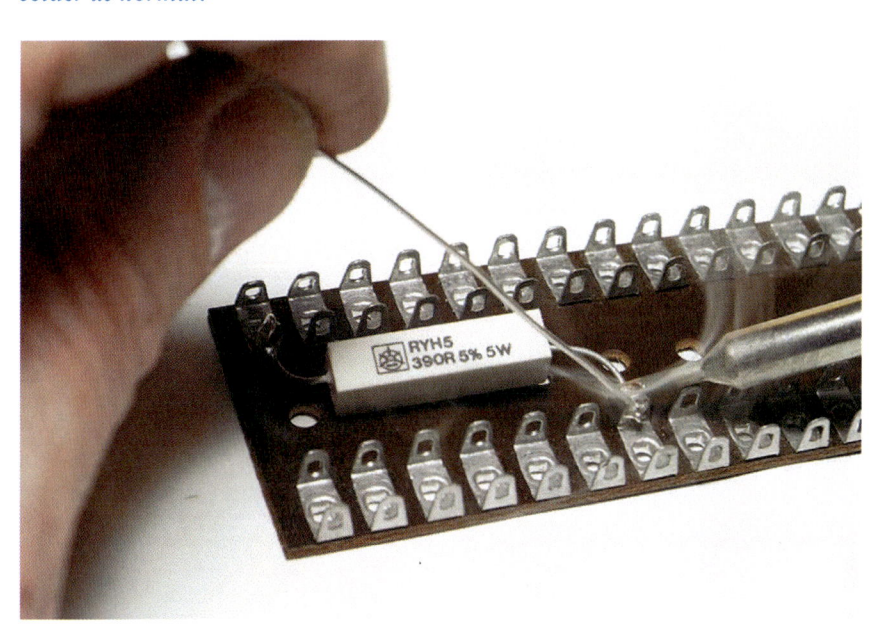

Don't be afraid to apply more heat with larger assemblies like these. They contain more metal than, say, an ordinary p.c.b so allow more time for the iron to heat the joint so that solder can flow properly.

Fatigue & Breakage

Earlier I explained how multi-stranded hookup wire is inherently flexible, which also makes it vibration proof – that's why a car's electrics are full of multi-stranded wiring, and test equipment probes use ultra-flexible multi-stranded wire for the same reason.

Solid-core wire can be bent and will stay in shape (called "plate wiring") but if it's repeatedly bent or vibrated then it may eventually break somewhere due to fatigue.

The same is true of wires that have been soldered or tinned. No longer is the wire 100% multi-cored and flexible – instead it's been turned into a single core wire at the point where it's been soldered. This is potentially a weak spot and could eventually fracture due to fatigue, if subjected to continued vibration (e.g. in a car engine bay or in motorised equipment).

In a lot of equipment problems can be avoided by adding strain reliefs of some sort, to stop the wire being flexed where it's been soldered. Heatshrink tubing, or a dab of hot-melt glue, are ways of taking the pressure off the joint and ensuring soldered wires won't snap off due to vibration.

One of the reasons that higher-quality equipment and cars etc. use crimp terminals and connectors is that crimped (as opposed to soldered) connections retain all the flexibility of multi-core stranded wire from end to end, avoiding problems of wires breaking off.

A **ferrule** is a very neat way of terminating multicored copper wires and preventing stray strands of wire from doing damage (eg shorting to other wiring or components nearby). Ferrules are narrow metal sleeves that can help when connecting wires into screw terminal blocks etc. (see photo). Simply slide a ferrule over the stripped wire and clamp the terminal block's screw onto it, and the screw will grip the ferrule. For production use, special crimping tools are used to fix ferrules onto the wires and tidy up the ends.

Instead of soldering wires, they can be terminated with ferrules prior to fitting to e.g. screw terminal blocks. This makes them vibration-proof and also prevents damage due to stray wire strands.

Desoldering techniques

By putting into practice the guidelines in the *Basic Soldering Guide Handbook*, there's no reason at all why you should not obtain perfect results and eliminate any potential problems. Hopefully the examples give you plenty of guidance to tackle various soldering projects with confidence.

There's no substitute for getting some hands-on experience though, so I'd repeat the advice to try assembling a simple high quality electronic kit or two, such as those produced by Velleman and see how you get on. Powering up your first project successfully is a great thrill as every electronics hobbyist knows, and this often leads on to much greater things.

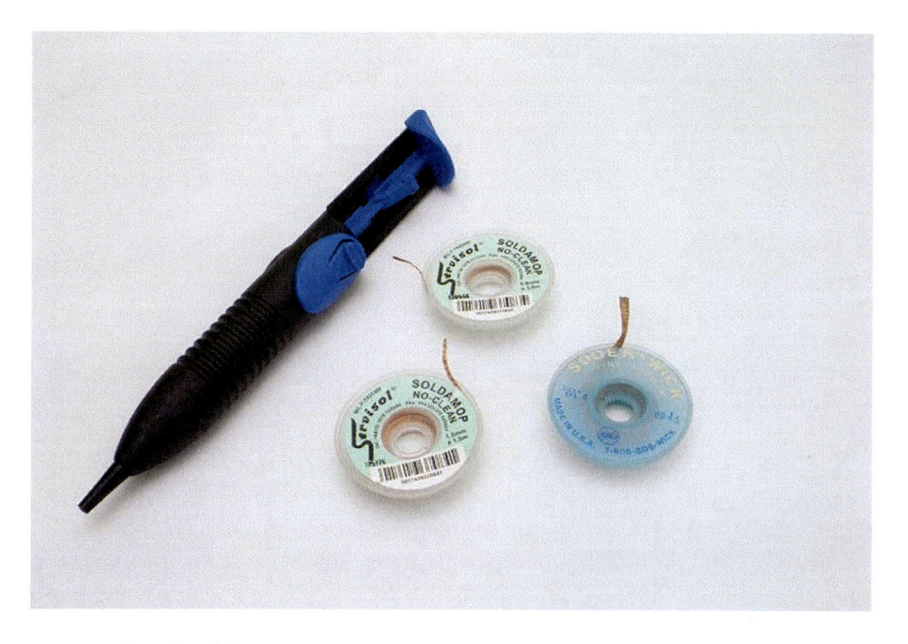

Popular desoldering products: a desoldering pump and various widths of desolder braid in handy dispenser reels. Both methods have their advantages in particular desoldering situations.

Let's now look at undoing the soldering procedure – what to do if things go wrong, or maybe you have to repair a circuit by replacing a faulty component.

A solder joint which is badly made is likely to be electrically "noisy", unreliable and will probably worsen over time. Expansion and contraction of the joint due to heating and cooling can also throw up intermittent problems later down the line. The joint may look OK but underneath it may have a poor electrical connection, or could work initially and then cause the equipment to fail at a later date!

These intermittent problems can be maddening to fix. TV repair technicians have an uncanny ability to go straight to a faulty solder joint because they see the same problem all the time, especially on equipment that has a "reputation" and experience tells them what needs repairing.

These dry or grey/ gray joints were caused by dirty parts, poor technique and inadequate solder coverage.

A solder joint that's poorly formed is called a "dry joint" (UK) or "cold-solder joint" or a "grey/ gray" joint (US). Usually it results from inadequate heating, dirt or grease preventing the solder from melting onto the parts properly, and is often noticeable because the solder tends not to wet the surface properly. Instead it forms beads or globules.

If it takes a long time for the solder to flow and wet the joint, that's another sign of contamination and that the joint may be a dry one, or the material is incompatible anyway. A matt, crystalline appearance points to inadequate heating: the solder cooled down too quickly and didn't flow properly. Alternatively the solder joint was disturbed while cooling down.

Whether you want to replace a faulty component or fix a dry or poor-quality solder joint, it's usually necessary to remove the troublesome solder and then re-solder it afresh. Naturally, there are tools and techniques that make the job easy. It's very bad practice to simply re-melt the joint and then lash out with the board, whiplash style, hoping that the molten solder will be flicked off the board.

The usual way of removing solder from a joint is to use a desoldering pump. A spring-loaded plunger is pressed down until it locks into position. It's released by pressing a button which sucks air back through a pointed nozzle, carrying any molten solder with it. It may take one or two attempts to clean up a joint this way, but a small desoldering pump is an invaluable tool especially for p.c.b. work and they are widely available at a very low cost these days.

A hobby desoldering pump primed for use.

Some desoldering pumps have a heatproof P.T.F.E. nozzle which may need replacing occasionally. Each time the button is pressed, a plunger clears the nozzle but solder particles and swarf are often ejected in the process; so when you prime the pump, point the nozzle into a small pot or old aerosol top to catch any debris that is spat out.

Remove the spout and clean out the pump from time to time. It's probably best to avoid the temptation of lubricating the plunger in case excess oil eventually contaminates the workpiece.

With very stubborn joints where the last traces of molten solder just can't be shifted, it sometimes helps as a last resort to actually *add more solder* and then desolder the whole lot again with a pump.

Suck up molten solder using a desoldering pump.

An alternative to a pump is to use desoldering braid which arrives in small dispenser reels. It's a flux-impregnated fine copper ribbon or braid which is applied to the molten joint, and the solder is then drawn up into the wick by capillary action. The solder rapidly cools and solidifies.

Desoldering braid is remarkably effective and for certain tasks, it can be more thorough than using a desoldering pump. I recommend that a small reel is bought (start with 1.5mm width) for the toolbox, to tackle larger or difficult joints which would take several attempts with a pump. Larger widths can be used to help with larger solder joints.

Care is needed when using desolder braid to avoid accidental damage so some practical techniques for using braid are explained next.

To use desolder braid, press the end of the braid down onto the joint using the tip of a hot iron, and let the solder melt underneath: the braid will then absorb the solder. The braid becomes hot so beware of burns. Once the solder's solidified on the braid, cut it off and discard.

The next photo sequence is a typical example of desoldering braid in action.

The Basic Soldering Guide Handbook

Above, desolder braid is versatile and is sometimes more effective than a desolder pump. It comes in small reels of various widths to suit the scale of work being tackled. This braid is 1.5mm wide.

Desolder braid can be used to remove excess solder, e.g. two i.c. pins shorted together. There is far too much solder here!

Above, press some desolder braid over the joint with a hot soldering iron, and the molten solder will be drawn up the braid. Take care to avoid damaging the circuit board due to overheating.

Remove the braid immediately and don't drag "whiskers" of molten solder around. The braid is snipped off ready for the next job.

Printed circuit boards can be damaged accidentally if the desolder braid is not removed soon enough. The excess solder soon solidifies, which solders the braid to the printed circuit board! A careless tug may then yank copper tracks or pads off the board. "Whiskers" of molten solder can also be dragged onto neighbouring pads unless the braid is removed carefully. So learn the techniques needed for using desolder braid swiftly and cleanly.

Whichever desoldering technique you use, whether a desoldering pump or braid it's not difficult to use so much heat that the adhesive holding the copper tracks onto the p.c.b. eventually fails, causing the copper track to lift away – everyone's worst nightmare (photo).

Typical copper track damage caused by overheating during soldering or desoldering. The track has lifted off.

If this happens, remove the iron immediately and allow the area to cool (a freezer aerosol is valuable at such times). If you're lucky, you can maybe repair the lifted track with Super Glue, or add "jumper wires" to bypass the damaged solder pad.

Why not practice using desolder braid or a desolder pump on an old circuit board or discarded electronics? Also try desoldering other components such as switches or potentiometers to see how effective desolder braid or a pump can be in various circumstances.

A Freezer aerosol offers rapid cooling where excess heat has been applied during soldering. Also used in circuit faultfinding to identify overheating parts: the hottest component(s) thaw the freezer spray first.

That concludes this primer on the basic techniques for soldering successfully. You have now seen, step by step, all the essential stages of correctly soldering a variety of electronic components, as well as how to desolder them again if you need to put things right.

Some useful summary checklists are given in Chapter 7, including:

- Quick Summary Guide – How to make the perfect solder joint
- Soldering Toolkit Checklist
- Troubleshooting Guide
- Summary of Potential Hazards and Basic First Aid.
- Useful web links
- Conclusion.

7 USEFUL CHECKLISTS

HOW TO MAKE THE PERFECT SOLDER JOINT

- Ensure materials to be soldered are compatible with tin/ lead or lead-free solder.
- All parts must be clean and free from dirt and contaminants.
- Try to secure the workpiece firmly during soldering.
- Brand new soldering iron tips must be flooded with solder immediately, the first time they are used.
- Wipe the tip of the hot soldering iron on a damp cellulose sponge at frequent intervals. Then "tin" the iron tip by applying a small amount of solder.
- Aim to heat all parts of the joint for typically under a second or so, to bring them up to the same temperature.
- Continue heating and apply sufficient rosin-core tin/ lead or lead-free solder to form a complete joint.
- It takes less than a second to solder the average p.c.b. joint. It should be smooth and shiny, and through-hole joints should be slightly convex in shape.
- Return the soldering iron to its stand after each operation.
- Do not move parts until the solder has cooled.
- Tin the soldering iron tip and clean it well, when switching it off, ready for next time.
- Consider using e.g. electronics flux dispenser pens or Colophony (rosin) to help with difficult joints.

SOLDERING TOOLKIT CHECKLIST

ESSENTIAL TOOLS
- Electric soldering iron, rated 25-40W
- Soldering iron stand and sponge
- 60/40 tin/lead solder or lead-free solder, 22SWG/ 21AWG for general use
- Pair of electronics/ "radio" long-nose pliers
- Pair of electronics/ "radio" wire cutters
- Pair of adjustable wire insulation strippers
- "Helping Hands" component holder
- Desoldering pump and/ or
- Desolder braid of various widths
-

GOOD TO HAVE
- Multicore TTC1 Tip Tinner & Cleaner
- Brass wool-type soldering bit cleaner
- Thicker gauge solder for larger joints
- Glass-fibre pencil abrasive cleaner and/ or
- Abrasive eraser block to clean metal
- A small selection of different 'bits' for your soldering iron
- Soldering tools (scraper, brush etc.)
- Benchtop fume absorber
- A magnifier loupe to inspect work
- Freezer aerosol
- PCB solvent cleaner aerosol

FOR ENTHUSIASTS
- Soldering iron station (digital display or analog control) with matching soldering iron and holder
- "Round-nose" pliers to form wire shapes
- "End" cutters for flush cutting of solder joints
- Automatic wire stripper tool
- Gas soldering iron for on-the-spot quick soldering tasks, heatshrinking tubing, and/ or:
- Electric hot air gun for heatshrinking
- Electric Solder Pot for tinning leads
- Solder Flux dispenser pen (eg Chemtronics)
- Circuit board holding frame

TROUBLESHOOTING GUIDE

Solder won't "take" (wet) and won't flow over the joint – molten solder forms beads or "ball bearings" instead of flowing properly.

Grease or contaminants present; material may not be suitable for soldering with standard lead/ tin or lead-free solder anyway, e.g chromium.
Treat contaminated parts with (abrasive) cleaners etc. as required to expose base metal. Some metals can't be soldered with electronics-grade solder.

Solder doesn't melt or flow very well – the joint is crystalline or grainy-looking - a grey "cold-solder" or "dry" joint.

Joint has been disturbed before being allowed to cool naturally, or the joint was not heated adequately. Too large a joint – too much metal present – and/ or the soldering iron temperature or power rating are too low.
Desolder and remake. Apply heat for a longer period, or use a higher power soldering iron, or check the temperature setting and raise it if possible.

Solder joint forms a "spike" and applying the iron again makes it even worse!

Probably overheated, burning away the flux. The iron, when removed, would cause the solder to stand up in a spike. It is usually best to desolder and remake the joint freshly again.

The copper foil of my p.c.b. has lifted off the circuit board!

Excessive use of heat has damaged the adhesive. Provided the track hasn't broken, it may be repairable with Super Glue, or re-wire the board with jumper wires.

Brown varnish-like deposits are left behind after I finish soldering.
These are the remains of rosin flux and are nothing to worry about.
The material can be removed with Isopropanol or PCB cleaner, if you want to tidy up the board and inspect your work.

POTENTIAL HAZARDS

It's extremely rare that soldering iron operators receive any burns or other injuries from the use of hot soldering irons. Soldering is perfectly safe provided that common sense precautions are taken during the soldering operation. Here are some of them:

- Especially if working on mains or high voltage equipment, it is safest not to work by yourself. Otherwise in the event of an accident there would be no-one nearby to help you. An RCD (GFCI in the USA) helps safeguard against electric shock.
- Components are very hot after soldering, so let them cool before handling them to avoid skin burns.
- Beware of splashes of molten solder caused by careless handling of a hot soldering iron. Consider wearing safety spectacles, especially at trainee or student level.
- Beware of energised components (capacitors, batteries etc.) being shorted by molten solder and ejecting solder splashes due to arcing.
- Always park a hot iron safely on a stand in between use — never place it down flat nor hang it vertically by the bench.
- Keep a hot soldering iron away from its mains cable (silicone cables reduce the risk of accidental damage).
- Beware of wire offcuts flying off (danger to eyesight) when snipping wires to length.
- Avoid inhaling solder and flux fumes as this can irritate the respiratory tracts, especially in sensitive cases (e.g. asthma sufferers).
- Some people may be allergic to Rosin (as used in flux). Rosin is used in some adhesives and e.g. band-aid/ sticking plaster adhesive, so if these types of product give you a skin rash, consider using non-allergenic gloves when soldering.
- Wash hands after handling solder, especially before consuming food.

BASIC FIRST AID

- In the event of a burn, cool the affected area immediately. Use plenty of cooler running water – but avoid ice cubes etc. as they can cause nerve damage after a time or inhibit the flow of blood to the affected area.

- Remove any objects which may be constrictive, before any swelling starts (rings, watches, bracelets etc.).

- Do not prick blisters nor apply ointments, salves or lotions at this stage. Seek medical attention for more serious burns, or ask a Pharmacist for advice in lesser cases.

- Eyesight problems are exceptionally rare, e.g. wire offcuts or solder splashes lodging in the eye area, and should be treated by a qualified first-aider or A&E. The best you can do is irrigate the affected area with e.g. a first-aid eyewash bottle or fresh water. Seek professional medical help straight away.

Commercial, industrial or educational users

NOTE THE LOCATION OF YOUR FIRST AID POINT HERE

_____

FIRST AID PHONE / EXTN. No.:

_____

USEFUL RESOURCES AND WEB LINKS

Details of the Antex range of soldering equipment, solder tips and spare parts as featured in this book can be obtained from **www.antex.co.uk**. UK and international distributors are also listed.

Suggested UK/ US/ Australian suppliers of soldering equipment, kits and components:

- ESR Ltd. (UK, soldering equipment, Velleman kits) http://www.esr.co.uk
- Bowood Electronics (UK, Antex irons and spares, electronic parts) http://www.bowood-electronics.co.uk
- Rapid Electronics (UK, components, tools, equipment) http://www.rapidelectronics.com
- Cricklewood Electronics (UK, CCTV, Antex soldering irons and spares, components) http://www.cricklewoodelectronics.com
- Maplin Electronics (UK electronics retailer and mail order) http://www.maplin.co.uk
- Farnell Electronic Components (Major UK electronics supplier) http://www.farnell.com
- RS (Major UK electronics supplier) http://uk.rs-online.com
- Brewsters Ltd (UK soldering equipment mail order specialists) http://www.soldering-shop.co.uk
- Velleman UK (electronic kits) http://www.velleman.co.uk
- Quasar Electronics (UK, Velleman kit mail order retailers) http://www.quasarelectronics.co.uk
- Digikey (USA supplier, export) http://www.digikey.com
- Mouser Electronics Inc. (USA supplier, export) http://www.mouser.com
- Jameco Electronics (USA supplier, export) http://www.jameco.com
- Jaycar (Australian supplier, kits, components, export) http://www.jaycar.com.au
- Altronics (Australian supplier, kits, components, export) http://www.altronics.com.au
- Kemo Kits (Germany, trade only) http://www.kemo-electronic.de
- Hobbytronics (UK hobby mail order supplier) http://www.hobbytronics.co.uk
- Multicore Solders (now a Henkel brand) http://www.henkel.com

CONCLUSION

Hopefully the *Basic Soldering Guide Handbook* will give you the confidence to try your hand at electronic soldering. It's really a lot easier than it sounds, and armed with the guide's advice and photographs, the next thing to do is invest in a decent-quality soldering iron such as the excellent range manufactured by Antex that will serve you well for years to come.

Soldering together a commercial electronics kit – such as the professional designs produced by Velleman or Jaycar – is a great way of testing and extending your new skills.

In the early stages, don't struggle with old, surplus or recycled electronics components which may be difficult to solder and cause disappointment. Start with a small, simple kit costing a few pounds or dollars: kits usually contain all the electronic components needed so all you need to do is follow the diagrams and solder the components together onto a printed circuit board supplied. There's every chance that a good quality kit will work first time and you will enhance your skills and boost your confidence at the same time. Practise the art of desoldering using some old circuit boards too.

Successfully completing your first few projects is immensely satisfying and, as every electronics hobbyist knows, after building your own project the initial power-on can be thrilling! With luck you'll catch the electronics "bug" and with experience you will gain more confidence, but avoid the temptation to tackle something too complex until you're ready to extend yourself further.

I really hope that you will be encouraged to explore the fascinating world of microelectronics further, just as I have since I was a schoolboy, and that you will enjoy acquiring these satisfying new skills.

Your feedback, comments and questions are welcomed by email to alan@epemag.demon.co.uk.

Now it's over to you! Good luck with your soldering.

Alan Winstanley

ALSO BY THE SAME AUTHOR

THE BASIC SOLDERING GUIDE (KINDLE® EDITION)

Now available for the Kindle® platform, the Basic Soldering Guide (B00E8NEGAA) includes the text and high-quality images to help you solder electronics successfully. Free reader software for PC, Mac, iPad and Android are available from Amazon. Please consult the Amazon site in your country for details.

AN INTRODUCTION TO GAS SOLDERING IRONS (KINDLE® EDITION)

This handy guide was written to fill in the many gaps in a typical gas iron's operating instructions, explaining the ins and outs of choosing and using a gas soldering iron, looking at how they work, the key differences between different types, information about butane gas and key safety considerations too. Several popular gas soldering irons are also reviewed and it's a must-read for anyone thinking of buying one.

The Introduction to Gas Soldering Irons is available on Amazon Kindle® (B00J4Y3NSY)

Free reader software for PC, Mac, iPad and Android are available from Amazon. Please consult the Amazon site in your country for details.

COMING SOON

THE BASIC SOLDERING GUIDE WORKLAB EDITION

A version specially designed for benchtop use by students, trainees or hobbyists. Check alanwinstanley.com for latest news.

ABOUT THE AUTHOR

Alan Winstanley is a UK-based freelance writer in technology and the Internet, and has been involved in electronic project design, tutorials and features since a schoolboy in the 1970s. Many electronics hobbyists and students will know of his work in EPE magazine (Everyday Practical Electronics magazine www.epemag.com), the UK's No. 1 hobby electronics magazine which has published his work for decades.

Alan has also held various positions in manufacturing industry at national and international level, in sales, product design and development and gained much experience in many different aspects of business, marketing and commerce over the years.

He also has an interest in macro and technical photography, gardening and horticulture and enjoy trekking on his mountain bike when the English weather permits.

Alan Winstanley

Printed in Great Britain
by Amazon